BIG AND BRIGHT   A History of the McDonald Observatory

D0901531

History of Science Series, No. 4

William Johnson McDonald (1844–1926)

BIG AND BRIGHT
A History of the McDonald Observatory

by David S. Evans and J. Derral Mulholland

University of Texas Press, Austin

First edition, 1986

Requests for permission to reproduce material from
this work should be sent to:
   Permissions
   University of Texas Press
   Box 7819
   Austin, Texas 78713-7819

Library of Congress Cataloging-in-Publication Data
Evans, David Stanley.
   Big and bright.
   (History of science series   ; no. 4)
   Includes index.
   1. McDonald Observatory—History.   2. Astronomical
observatories—Texas—Fort Davis—History.   I. Mulholland,
J. Derral (John Derral), 1934–   II. Title.
III. Series.
QB82.U62F674   1986   522'.19764'934   86-7044
ISBN 0-292-70759-2

*DEDICATED TO the achievements
of our predecessors and our colleagues,
and especially to FRANK N. EDMONDS, JR.,
who led the way in Austin*

Contents

Photo section following page 60.

Preface

Roughly fifty years ago, a remarkable collaboration was undertaken by two major universities, one a prestigious private school, the other a large southwestern state institution. One had astronomers; the other had money for astronomy. Despite mutual apprehension, a thirty-year compact was made for a joint venture in a major observatory.

Today, such a compact does not seem remarkable. These last three decades have seen the formation of a number of consortia for the joint operation of scientific research facilities. The astronomically aware reader will recognize AURA, USRA, CERN, ESO as acronyms for such groups. In 1932, however, rugged individualism was still the rule in science and academia, as elsewhere. In this case, though, two parties each had something the other needed badly. Vision overcame tradition and prejudice on both sides, and McDonald Observatory was born. One can see in its founding some of the roots of consortium "big science," and as such this event had a meaning much larger than just another big telescope. It presaged the future.

The collaboration between the Universities of Texas and Chicago gave rise to several attempts to establish a wider participation in astronomical cooperation, based at the McDonald Observatory. These efforts were evidently premature, but by any measure the bilateral experiment was a success. As we approach the golden anniversary of the first astronomical activity on Mt. Locke, it seems an auspicious time to recount the origins of McDonald Observatory, the considerable birth pangs and considerable successes, the growing self-sufficiency of Texas astronomy. Although the more recent past and the future are given enough coverage to show that we have not been idle, this is essentially a history of the original McDonald telescope. The climax of the detailed narrative is the dedication and entrance into service of what is presently the McDonald Observatory's largest telescope, the 107-inch reflector. To avoid passing judgment on ourselves, we leave the history of the present to be written by the historians of the future, who we hope will be better equipped to place it in its proper dispassionate context.

When we first conceived the project of writing the history of McDonald Observatory, just over a half a century after the original legacy, we naïvely perceived a danger that it might degenerate into a kind of continuous paean, an astronomical Hallelujah Chorus. Now that the task is complete, we find it containing passages more reminiscent of the Book of Job, and other times resembling a scene from *Animal House*. We have been made to remember that astronomers too are human, with all the human errors and weaknesses, but who still from time to time rise to almost superhuman aspirations and achievements. We do not dwell morbidly on the faults and frailties, but as conscientious historians neither have we omitted them where they are essential to our narrative.

We should not be able to write with such fidelity to the facts were it not for the generous assistance of many individuals who have shared with us their memories and souvenirs. Many have written letters, recorded tapes, lent photographs and even movies. We wish to express our appreciation to all, but especially to Frank Berkopec, William P. Bidelman, Bart J. Bok, Margaret E. Burbidge, Subrahmanyan Chandrasekhar, Ruth Lundin Douglas, Frank N. Edmonds, Jr., Frank K. Edmondson, Hildegard and Rizer Everett, Ruth Fuller, A. D. Glover, Jesse L. Greenstein, George and Linda Grubb, Eloisa and Tommy Hartnett, W. A. Hiltner, Dorothy Hinds, Lewis M. Hobbs, Paul Jose, Fritz Kahl, Marlyn Krebs, Nicholas U. Mayall, William J. McDonald II, Violet Locke McIvor, Keesey and Lillian ("Bit") Miller, Lucy Miller, Richard I. Mitchell, William W. Morgan, R. Edward Nather, John S. Neff, John M. Neilson, John A. O'Keefe, Thornton L. Page, Jubal Parten, Daniel M. Popper, E. J. Prouse, Franklin E. Roach, Sarah Kuiper Roth, Paul M. Rybski, Ronald A. Schorn, Walter Steiger, Pol Swings, Jean Texereau, Robert G. Tull, Peter van de Kamp, Peter O. Vandervoort, Gerard de Vaucouleurs, Eddie Webster, W. G. Whaley, and Ewen A. Whittaker.

We are especially grateful for the assistance and facilities afforded to us by Harlan J. Smith, and Curtis D. Laughlin, respectively director and superintendent of the McDonald Observatory, for access to archival material. In addition to their official help, they have made important contributions to the text, not just by their suggestions, but also by their hospitality; significant portions have been written while the junior author was lodged in House A, far from the broiling heat of an Austin summer. Our thanks are also due and rendered to James N. Douglas, director of the University of Texas Radio Astronomy Observatory; Paul A. Vanden Bout, head of the University of Texas Millimeter Wave Observatory; and Frank N. Bash, chairman of the Astronomy Department.

We are deeply indebted to the Barker Historical Collection of the University of Texas at Austin, to the Freiberger Library of Case Western Reserve University, and to the Yerkes Observatory for access to historical documents. Librarians Susan Hansen at Case Western Reserve and Judy Lola at Yerkes provided invaluable aid to our research. Mary Dutchover and Jane Wiant unearthed a collection of photographs and newspaper clippings dating back to 1939 in the McDonald Visitors Center. Arthur Dilly, executive secretary of the University of Texas Board of Regents, provided useful data from the regents' files. By permission of Spencer R. Weart and the American Institute of Physics, we have examined a transcript of an oral history recorded for AIP by Frank K. Edmondson, which provided valuable confirmation of some points derived from other sources. David DeVorkin, of the Smithsonian Institution's National Air and Space Museum, generously shared prior to publication his research on Yerkes Observatory during the years of World War II.

Sections of the manuscript have been read critically and commented upon by Frank Bash, Anita and William Cochran, Frank Edmonds, Hildegard and Rizer Everett, Jesse Greenstein, Ed Nather, Donald Osterbrock, Jean Texereau, Paul Vanden Bout, and Gerard de Vaucouleurs. Despite a heavy schedule of other commitments, Harlan Smith has examined every word of the manuscript and offered many helpful suggestions. Equally valuable was the detailed scrutiny of Ann S. Parish, who checked the manuscript for linguistic usage. Each of these readers has improved the quality of the final text, for which we accord them our enthusiastic thanks.

Last, but far from least, we gratefully acknowledge the cheerful efforts of Mrs. Alice Herzog, for her extensive help both in typing the first draft manuscript and then entering it into the word processor adopted for subsequent revisions. Her lot was not an easy one. Someday, she may forgive us.

Given the frailty of the human mind and hand, we have surely omitted to mention some important debts. For such lapses as exist, we apologize sincerely. They result not from meanness, but rather from our inability to deal adequately with the overwhelming torrent of help that we have received.

Finally, we must call attention to the fact that, although both of us have for many years (nearly two decades in one case, over a dozen years in the other) done most of our astronomy in the McDonald Observatory and we have had massive support and encouragement from our colleagues there, this work is in no way an "official" history. We have felt perfectly at liberty to reject even the most well-intended advice, included Harlan Smith's, when it seemed appropriate. We have even on occasion rejected one another's opinions, in temporary disagreements that suggest that literary collaboration is occasionally somewhat akin to civil war. We have tried our best to weave a correct and coherent story from hundreds of hours of research on both original documents and first-hand reminiscences. The opinions and interpretations presented here do not necessarily represent the views of the University of Texas, the McDonald Observatory, or anyone other than the authors. The fault is entirely ours, if any there be.

D.S.E.
J.D.M.

BIG AND BRIGHT   A History of the McDonald Observatory

The stars at night are big and bright,
  Deep in the heart of Texas.
The prairie sky is wide and high,
  Deep in the heart of Texas.
The sage in bloom is like perfume,
  Deep in the heart of Texas.
Reminds me of the one I love,
  Deep in the heart of Texas.

CHAPTER 1

The Tourist's Observatory

The white domes shine in the sunlight against the deep blue of the West Texas sky. Across the Rio Grande, a hundred miles away, Mexican mountaineers use them as landmarks; for them, Mt. Locke is *la montaña con dos huevos,* the mountain with two eggs.[1] As far as the eye can see, cliffs and boulders, mesas, and conical peaks repeat themselves diminuendo into a hard purple skyline. A brace of deer pick their way along the slopes under the scrubby pines and liveoaks, as a red-tailed hawk cavorts on the thermals along the russet cliffside. It is afternoon at McDonald Observatory.

Just below the final rise of the mountain, cars pull in and out of the W. L. Moody Jr. Visitor Center. Above, in the biggest of several domes, a guide describes the 107-inch telescope to a small group, some of the more than 40,000 tourists who visit this place every year. No one visits here on a spur-of-the-moment whim to blow a few minutes to the wind. They have to come on purpose. It is an 80-mile detour from the San Antonio/Dallas–El Paso interstate highway I-10/I-20, only an extra 55 miles for those following U.S. 90 between El Paso and San Antonio. Still, they come at an average rate of 100 per day.

Nothing could be more deceptive than the almost sabbath calm that pervades the mountain. There is no hint of scientific activity nor of the immensely complicated administrative and technical procedures needed to bring astronomers to their rendezvous at McDonald, always from hundreds, often from thousands of miles away. Sometimes the timing of the rendezvous is crucial for the observation of an event that will occur only at a certain moment, perhaps not again for decades. The observer must find the allocated telescope in good repair, ancillary equipment properly installed and working, data acquisition systems and computer programs free of defects. And all of this must be done in the splendid isolation of an arid mountaintop nearly 7,000 feet high, 16 miles from the nearest town, 40 miles from the nearest airfield or railroad.

Most of this is achieved behind the scenes. The tourists do not see the instrument development and construction teams back in Austin, 450 miles to the east. Neither do they glimpse the complex task of allocation of observing time, with planners playing musical telescopes with the fifty or so astronomers who wish to use them. Transport arrangements must get them to this remote spot, feed them, house them. The local maintenance crews are unseen, busy in their workshops, keeping everything running. Occasionally, a visitor might see a strange cutoff truck with a jury-crane go by, as the operations crew goes to install a piece of equipment on one of the telescopes. A passing utility vehicle carries a working party from the physical plant, which not only provides all the usual municipal services to this village of over a hundred souls, but also helps to provide technical services at a standard usually found only in large cities.

There are few astronomers to be seen. With rare exceptions, they are just waking from the few hours of sleep that they can manage after having left their domes at dawn. Over coffee, they look at the night's results while waiting for their first meal of the day. To help them withstand the night watch, in winter fourteen hours long and often bitterly cold, a normal dinner will be supplemented with a midnight lunch. Optical observers are night people.

Even when the night begins, there is not much for an onlooker to see. Astronomers abhor extraneous light. One might glimpse a night assistant, controlling the motions of the telescope from a dimly lit control panel. The astronomer might be standing with one eye glued to the guiding eyepiece, making minor corrections to keep the target centered on the crosshairs. Increasingly in this electronic age, he might instead be hidden away in a separate room, monitoring and controlling the torrent of data pouring into the memory of a computer. Noticeable signs of activity erupt only when service staff have to rouse from their beds to repair some malfunction.

At the end of the observing run, the astronomer descends to tiny Marfa Municipal Airport, where World War II B-25 bombers from the nearby flight school shot practice landings when the observatory was new.[2] There, a twice-weekly charter awaits to take astronomers, students, and engineers, along with the decks of computer cards, printed listings, magnetic tapes, or photographic plates containing the data, back to Austin. Returned to the daytime schedule of ordinary people, the observer will spend many hours or months interpreting the data from the few nights of looking through the narrow open slit in the domed roof. Eventually, those data will be incorporated into a final product—published reports of new knowledge about the Universe and its contents.

Thus passes the life of an astronomer. It is physically and intellectually demanding, leaving little time for the ordinary everyday pursuits.

This is McDonald Observatory, but the description would not be very different for most others. McDonald, one of the world's largest university-operated observatories, has changed considerably since it was conceived over fifty years ago. The photographic plate was king then, and all plans were geared to it. Photography remains important, but the computer revolution and the rise of microelectronics have changed astronomy beyond the wildest dreams of the founders. Even so, observatories are telescopes, and the telescopes continue to play their historic role as collectors of light, no matter what method is used to record it. They will do so for decades into the future.

If the scientific fathers of McDonald Observatory might be surprised at what has happened to their legacy, how much greater would be the astonishment of the benefactor, could he see what has been done with his money after fifty years.

William Johnson McDonald died on 6 February 1926, an introverted, unmarried, and prosperous banker in Northeast Texas. Unknown to most people, McDonald had had a "bee in his bonnet." This Scots phrase implies a fixation on a somewhat eccentric idea. The Paris (Texas) banker had a bee in his bonnet that continues to buzz around West Texas today. To his family's chagrin and everyone's surprise, he bequeathed the bulk of his large fortune, over a million 1926 dollars, to the regents of the University of Texas, ". . . to be used and devoted . . . for the purpose of aiding in erecting and equipping an astronomical observatory to be kept and used with and as a part of the University for the study and promotion of astronomical science."

The closest of his kin were accorded inheritances of $15,000 each, equivalent to almost $100,000 in today's values. Most of them were not satisfied. The disinherited felt even worse. It seemed clear to the disappointed family that the deceased had taken leave of his wits before taking leave of this world. They sued to upset the will, hiring both a smart local attorney and a distant relative who practiced law in Houston. The legal maneuverings are described later, but first we should take a closer look at William Johnson McDonald (1844–1926). On examination, he turns out to have been a shrewd and successful businessman, hardly of infirm mind.

CHAPTER 2

The Benefactor

In the late summer of 1837, a train of wagons from Tennessee moved through Arkansas to the new Republic of Texas. Its leader, John Johnson, had selected a site for himself on the Red River the previous year. He was now bringing his wife and ten of their eleven daughters to homestead it. The travelers settled in what is now Lamar County and began building homes and clearing land, always vigilant to repel Indian attacks.

In that same year, Dr. Henry Graham McDonald resigned his position as government physician to the Choctaw Indians after their forcible removal from their ancestral lands in Mississippi to the Oklahoma Territory. Like Johnson, he was of Scottish descent, and like many of his contemporaries, he turned his eyes toward Texas. There, as one of the rare certified medical specialists on the frontier, he established a successful practice by riding a regular circuit to visit his farflung clients. He began to invest his earnings in local real estate.

After an interval of successful combat against the virgin land, John Johnson brought his family into the little town of Pin Hook, now called Paris, the seat of Lamar County. He built a hotel there, leaving an overseer to look after his farms. Dr. McDonald often stayed at the hotel during his medical circuit journeys, and he eventually fell in love with the proprietor's eldest daughter, Mrs. Sarah Turner, a widow with two young children. They were married on 8 February 1844. Their first son, William Johnson McDonald, was born on 21 December of that same year, just over a year before the Republic was integrated into the United States of America. Henry Dearborn McDonald was born in 1847 and James Thomas McDonald in 1850.

The Life and Death of William Johnson McDonald[3]

The new husband and father continued to add to his holdings, eventually becoming one of the major landowners in the county. Sarah died in childbirth in 1852, and the daughter that she was carrying outlived her by only two months. The sons that she left behind were then aged two, five, and eight years. The three brothers remained very close throughout their lives.

A neighbor was employed as housekeeper to care for the boys, and Elizabeth Roberts eventually became the second and last of the doctor's wives. Relations between the boys and their stepmother were not cordial, and they spent much of their time at the home of their mother's favorite sister, their aunt Nancy Wright. The boys were educated at a school in nearby Howland until after the death of their father in 1860.

The boys were placed under the guardianship of the Reverend J. W. P. McKenzie, who had met their father while a Methodist missionary to the Choctaw. When his health began to fail, he came to Texas, where he founded a college in 1841. Consequently, the McDonald brothers transferred their educational pursuits to the McKenzie Institute at Clarksville.

In its prime, this institution attracted students from several of the neighboring states as well as from Texas. The curriculum was in the classical Scottish style, including Greek, Latin, mathematics, philosophy, and Bible study. Living conditions were austere. Hefty, red-haired William McDonald appeared to find his niche. An outstanding student, he gobbled up all the knowledge he could, taking a special interest in nature subjects. His studies were interrupted by a short stint in the Confederate forces at Pittsburg in East Texas. He saw little or no action and on his return completed his studies in 1867.

The war left the South impoverished, and it was a hard time to find a job. In an effort to survive, William turned to a variety of occupations, including the trade of printer. Finally, he was apprenticed to a law firm in the small town of Mount Pleasant, Texas. He learned enough about the law there that eventually he returned to Clarksville, seat of Red River County, where he opened a law office in partnership with Marshall L. Simms. Fees from his practice in civil law enabled him to make loans. During the economic depression of the 1870's, he also bought up Red River County warrants at about ten cents on the dollar. That paper was eventually worth its full face value again. At the end of the depression, the economic recovery made William Johnson McDonald a rich man.

Depression moved to boom time, and the spirit of the time was such that McDonald abandoned the law for banking. The free-wheeling atmosphere that characterized banking in that period is perhaps best compared with that of the automobile industry in the 1910's or the electronics business in the 1960's. Small local concerns sprouted everywhere like crabgrass. William McDonald moved in with characteristic energy. Supported by his two brothers, he was founder and president of the Citizens Bank of Clarksville in 1885, the First National Bank of Paris in 1887, and what would become the First National Bank at Cooper in 1889. In 1889 also, his first bank was merged with the First National Bank of Clarksville. Consolidate money with money. McDonald was often known to say that the original stake was all he ever made for himself. "The rest was merely money which that $5,000 made for me."[4]

In 1887, McDonald moved to Paris (Texas), where he resided for the remainder of his life. For many of these years, he occupied a modest furnished room on the second floor of a building owned by cabinetmaker S. W. Wilson. The shop was on the ground floor, and the owner and his wife shared the upper level with their wealthy tenant. At about this time, McDonald's imagination was fired by Camille Flammarion's book *Popular Astronomy,* recently acquired by his sister-in-law Irene, Henry's wife. William acquired a small telescope, and he and the cabinetmaker sometimes observed the heavens from the backyard.

Indeed, with increasing wealth, McDonald was able to indulge himself in various ways. He learned to play the flute. His library, now housed at the McDonald Observatory, shows an astonishing breadth of interests. All of the books are "serious," ranging from the classics of English literature to a large selection of works on botany, zoology, geology, history, and items of Texana—accounts of local geography, nature study, and history. The collection includes what must then have been daring purchases in the fundamentalist South, the works of Charles Darwin and Thomas Henry Huxley.[5] McDonald could, and frequently did, quote classic poets from memory.

He was able to travel, making three journeys to Europe and one to Mexico, as well as several within the United States. He attended the summer school at Harvard in 1895 and 1896 (where he described himself as a "farmer"), taking courses in botany and visiting the Harvard College Observatory. A skilled and impassioned amateur botanist, the increasingly wealthy banker trekked to Scotland to see the heather in bloom and to northern Canada to view rare wildflowers. Despite rustic origins and surroundings, William Johnson McDonald was evidently an urbane and intellectual man.

In 1916, a fire devastated Paris. McDonald lost most of his personal belongings, although the valuables in the fireproof vault of the bank were saved. He then moved three miles east of Paris to one of his farms, Yam Hill. For the next five years, a tenant looked after the farm, while the now-retired bank president went daily to his office by buggy. He eventually acquired a Model T Ford, but only used it once.

Before his death in 1983, the banker's grandnephew, also William Johnson (Bill) McDonald, recalled buggy rides from this period. The older man spoke quietly of the trees and flowers that he so loved. His kindness and his moderation, however, apparently did not dull his taste for precision even in minor matters. Asked on one occasion for money to enable young William to visit the State Fair in Dallas, he became cross because the request was not accompanied by a precise estimate of the cost.[6]

The end of the old banker's life was lonely. Brother James had died in 1905. After an unsuccessful bid for the governorship of Texas, Brother Henry practiced law in distant Corpus Christi for a while. He eventually returned to Paris, where he shared both house and office with William, but he shot himself in the head in 1925. William McDonald had no other close associates, and there is no record of any love affair.

The end was painful, too. McDonald developed a series of tenacious illnesses, including infections of the kidney, prostatitis (possibly malignant), and various degenerative conditions. He visited the Mayo Clinic and a New Orleans hospital in search of a cure for these afflictions. As a further aggravation, his sight had so deteriorated that he sometimes mistook the identities of business associates on the street. Late in 1925, his condition was diagnosed as terminal, and he was confined to the Sanitarium at Paris, Texas, to live out his last days. There, for three months, young William's mother, Leila, would come to read to him. He was attended constantly, by a male nurse at night and a female nurse by day. They later had tales to tell about how his mind would wander a little. At the time, however, Leila McDonald said that his mind was "as clear as a bell,"[7] and his attending physicians thought him mentally alert to the end.

William Johnson McDonald died on 6 February 1926, at the age of eighty-one years. Almost no one, maybe not even he, knew the extent of his fortune. Few enough knew what he intended for it.

The Will: A Million for Astronomy

McDonald had made his last will and testament on 8 May 1925, properly witnessed by two of his Paris bank employees. When it was opened and read a few days after his death, it proved a shock both to relatives and to the University of Texas.

A complete understanding of the complicated and sometimes bitter legal contests that followed would require a careful reading of the entire text of the document. It is reproduced as an appendix to this book, but for the present purposes, a brief summary will suffice.

The executors were designated as Morris Fleming, cashier of the Paris bank, and the First National Bank of Clarksville, for which Eugene W. Bowers, the cashier of the bank, would act. When finally probated on 21 March 1929, the estate totaled somewhat more than $1.26 million, of which roughly $150,000 was in cash, as compared with only $40,000 in real estate and $112,000 in notes receivable. More than $1.1 million was listed as "personal property," including the cash. A category listed curiously as "household goods" included two gold watches and one silver watch valued at $25 each, and a Colt revolver listed at $15. The weapon would acquire significance during the trials.

In very detailed instructions, the deceased banker specified the expenditure of $1,500 for his gravesite and monument and the bequeathal of $15,000 each to eight of his kin. Florence McDonald Rodgers and Lillian McDonald Brinton were nieces, daughters of brother Henry; William and Irene Stromeyer were minor children of Mrs. Brinton by a former marriage. Nephew Charles McDonald, one of the two sons of brother James, received only a life interest in U.S. Treasury bonds, with reversion to his children James, William, and Rosemary McDonald. In addition, they each received their own bequest of Treasury bonds.

The benefactor did not deal evenly with all his kindred. He left nothing to Charles' brother, Dr. William McDonald, apparently regarding him as having brought disgrace on the family through divorce. Nor did he leave anything to the family of his already-deceased half-brother Dan Taylor, nor to the family members of any of his numerous aunts.

The clause that raised eyebrows, triggered five consecutive court actions to overturn the will, and made this history worth writing is contained in section 5:

> All the rest, residue and remainder of my estate, I give, devise and bequeath to the Regents of the University of Texas, in trust, to be used and devoted by said Regents for the purpose of aiding in erecting and equipping an Astronomical Observatory to be kept and used in connection with and as a part of the University for the study and promotion of the study of Astronomical Science. This bequest is to be known as the W. J. McDonald Observatory Fund. As soon as my Executors shall have paid off all charges against my estate, settled with all legatees and have the estate in shape to turn over the residue to the Regents, they shall do so using only such time to accomplish this as in their judgment is necessary, reasonable and proper.
>
> Upon receipt of such residue the Regents shall proceed at such time and manner as in their judgment may seem best to execute the trust and they shall have full power and authority to administer and handle the same for the purpose of carrying out its purpose and object, and in so doing they may apply all or any part of the income from the bequest and all or any part of the corpus of the bequest as they may deem proper to use towards such erecting and equipping the Observatory, it being my intention that the Regents shall have full power and authority to handle, use and appropriate this bequest, corpus and all accrued income in such manner as to them may seem best in order to carry out the object of the bequest, the only limitation on their authority and power being that the bequest is intended solely for the use and benefit of an Astronomical Observatory in one or all the ways hereinbefore mentioned.

An abbreviated version of the first sentence is painted on the wall of one of the ground floor rooms of the 82-inch telescope building at McDonald Observatory.

The university authorities learned the news by a telephone call from a reporter, who read the wire service report that had just come off the teletype. Dean of Science Harry Yandell Benedict called it "like lightning out of a clear sky."[8] He and his thunderstruck colleagues soon found that they had to cope with new friends and legal foes. Within days, they received a spate of congratulations, of letters of advice, and of offers of land and facilities suitable as sites for the observatory. Quickest off the mark were several Chambers of Commerce in West Texas, some of whose letters were dated within three or four days of the publication of the will.

The residue amounted to rather more than one million dollars—equivalent to over six million in 1986 buying power. A million in tribute for astronomy, with a few modest leavings for kith and kin. With one notable exception, the relatives were malcontent. Not only did seven of the eight legatees want more, but they were joined by many of those who were left out. The deceased millionaire had taken no testamentary notice of the large family of his half-brother, Dan Taylor, nor of various grandnephews and grandnieces. They would jointly contest the will, on the grounds that the will's author was "not of sound mind and disposing memory" at the time of its drafting.

Nephew Charles McDonald did not participate in the challenges, consistently saying that his Uncle Will had always been perfectly sane. He would be content with his $15,000.

Before being settled, the cause was to appear in court five times. The various suits and appeals were defended by the legal executors, assisted by the university and the staff of the Attorney General of the State of Texas. Consequently, the legal records appear under the obscure title of *Mrs. Florence Rodgers et al. vs. Morris Fleming et al.*

CHAPTER 3
The Legal Contests

The University's Response

The university, of course, was inclined to agree with Charles McDonald. The deceased had been "of sound and disposing memory," but it was going to be necessary to convince a jury of that. Dean Benedict had earned an astronomical doctorate at Harvard University. He had forsaken active research to undertake a career as an educator, but he was well equipped to represent the university's case. Thus, President Walter Splawn designated him to be the principal activist on the academic side.

Among McDonald Observatory astronomers, there is a persistent tradition that a desire to endow astronomy was the ground for the allegation of insanity. According to this tale, reputable persons were found to testify that such an ambition was not in itself prima facie evidence of dementia. Charming though this story may be, it exaggerates a bit. There were indeed real fears that this was to be the tack taken by the disaffected family. Benedict wrote that "If Mr. McDonald had left this money to a hospital, many citizens would have understood and approved his motives. The attorneys will argue that a gift of this size for so remote a subject is evidence of some mental queerness."[9] Benedict explored possibilities for the calling of expert witnesses. He finally chose Professor D. W. Morehouse, then famous for the comet named after him. In the end, Morehouse could not appear. In fact, no astronomer was sworn as a witness at any of the judicial proceedings.

Benedict wrote to the directors of several observatories that had been founded or supported by legacies, seeking documentary support to demonstrate that this was a common practice among wealthy men desirous of perpetuating their memories through the endowment of research, and that astronomy was beneficiary of a steady stream of such gifts.

One of the recipients of Benedict's letter was Edwin B. Frost, Director of the Yerkes Observatory of the University of Chicago. Charles T. Yerkes had been a successful Chicago businessman in the 1890's, and the idea of having the world's largest telescope named after him appealed to his spirit. Yerkes did not wait to die; he gave the university enough money to start the observatory while he was still living, augmenting it with a bequest on his death. The Yerkes telescope, still in service and still a wonder, is probably the largest refracting telescope the world will ever see. Its primary lens is 1 meter (40 inches) in diameter and was figured by the elder C. A. R. Lundin, whose son figures later in our story.

Frost's effectiveness was waning, because of advancing blindness, but his memory was good and his head clear. He was later to make an all-important decision that would affect the McDonald Observatory's entire future, and his response to Benedict's plea for help left nothing to be desired.

Frost's Defense of Astronomical Bequests

Frost began his letter with the observation ". . . observatories founded by level-headed, successful men have constituted memorials more widely known and more significant than those of almost any other nature." Warming to the topic of the "great practical service that astronomy has done in the world's business," he cited several examples. "The practical men of the Great Plains must realize that the legal location of every acre of their land depends upon a parallel of latitude laid out by observation of the stars, and upon a meridian similarly determined." Then, ". . . every train runs safely because accurate time comes to each station . . . from someone who has observed the stars a few hours previously." And more, ". . . every pound of imported goods is safely brought to port by . . . skilled navigation . . . only possible by astronomical observation."[10]

Add aircraft to the ships and trains, and the arguments remain valid and even more forceful today.

Frost followed the commercial with a list of American observatories with private endowments. It is truly impressive. Besides Yerkes, the prominent names from the business and political world include Lick, Thaw, Frick, Mellon, Chabot, Longworth, Shattuck, Field, Dudley, Chamberlin, McCormick, Litchfield,

Ladd, Sproul, Washburn, Swasey, Lowell, Carnegie, Hooker, and Perkins. The cities, colleges, and universities that had benefited from their largesse include Oakland, Cincinnati, Albany, Denver, Dartmouth College, Williams College, Hamilton College, Swarthmore College, University of Chicago, University of California, University of Virginia, Brown University, University of Wisconsin, Dennison University, Yale University, Harvard University, and Ohio Wesleyan University.

Frost dwelt at some length on the connection between the Sun and our weather, and on the benefits to be gained from a better understanding of solar physics: "When such knowledge is available, it may be worth billions of dollars to the farmers and the businessmen and to all citizens of the country." There was a special word for Texans: "Every citizen of Texas is proud that his State can produce helium. Do they remember that helium was discovered by a French scientist, studying an eclipse of the Sun in 1868, and was not found on Earth until 27 years later?" [In fact, history now credits Pierre-Jules-César Janssen and Englishman Joseph Norman Lockyer with near-simultaneous independent discoveries.]

Frost's letter does not seem to have been used directly in the defense, but Benedict was greatly buoyed by it, and it prepared the psychological groundwork for a future collaboration.

The Trial Begins[11]

The case came up for trial before Judge Newman Phillips in the District Court in Paris, Texas, on 30 August 1926. As soon as it became clear that the will would be contested, the university regents sent one of their number, Sam Neathery, to the scene of the action. Apparently, neither President Splawn nor Dean Benedict attended. Indeed, Benedict wrote to a correspondent that he was not sure that he should, as his "views on religion were somewhat unorthodox."[12] As it turned out, the case was tried entirely on the issue of testamentary capacity, the merits or otherwise of astronomy as a science being mentioned only briefly in speeches by counsel. The law in such cases seems to be the following: a testator has the liberty to leave his property to whomsoever he wishes, with some restriction as to descendants and other kin under Texas law. To make a valid will the testator must understand the nature of the property that he has to leave and, at the time of making his will, he must be capable of understanding the matter in which he is engaged. He must be sane in the sense that he is not prey to delusions of things and circumstances that do not exist, insofar as they affect his testamentary actions. Insane delusions are defined as fixed ideas of things that do not exist and that are persistent and incapable of being removed by rational argument. A testator is entitled to be rationally misinformed and to give rein to prejudices, whether or not agreeable or reasonable to others. A will would be in doubt, however, if a disposition appeared to be the result of undue systematic persuasion by an interested party, or if the testator's judgment were overwhelmed by strong emotion or diminished by disease.

The establishment or contradiction of these principles was the goal of a cloud of some forty witnesses who deposed or testified at the original trial.

After two days of legal mechanics, the long saga of contradictory testimony began. Henry's widow Irene McDonald complained that her brother-in-law had persuaded them to return to Paris to help with the business and to care for him. He was, she said, insomniac, excitable, irritable, and prone to eccentric opinions. One of these was a hostility to the University of Texas—especially the Law School. At the time when he was paying grandnephew Jim's expenses there, he was alleged to believe that "it is no place to educate girls and boys at taxpayers' expense."[13] Irene McDonald testified that she knew that he was of unsound mind, despite her subsequent remarks to the contrary. Irene also testified that her brother-in-law feared that nephew William, the divorced scapegrace, would murder him. The pistol in his household was for protection.

In fact, there was no want of adverse witnesses. McDonald's personal secretary, Constance McCuistion, described his increasing weakness and the decay of his mental faculties during the previous five years. Others spoke of his loss of capacity, forgetfulness, endeavors to collect the same note twice, failure to recognize close associates in the street, and so forth. On the other hand, a score of witnesses testified to the soundness of his business sense even at the end of his life. When nurse Clarence Mills testified that McDonald objected to the sterilization of his catheters because "it wore them out," the defense countered with Fred Manton, the mortician called after Henry McDonald's suicide, which the defense used as a watershed event in the latter days of William's life. During the trauma of that day, 22 April 1925, W. J. was nervous and excited, but quite clearheaded. He gave detailed and exact instructions for the arrangements of his brother's funeral.

Some of the testimony was picturesque. Autrey Burnett operated a large barber shop on Lamar Avenue, where W. J. McDonald often went to be shaved. One day, after another barber had shaved him, he called Burnett into the back room "where the baths were" (barbershops commonly offered hot baths weekly to customers with no home facilities) and started to talk about astronomy. It was the coming science, the great wonder of the world, though neglected in the universities. All it needed was a bit of money. "If we had a big enough telescope, we could see into the gates of Heaven and see who was there."

Beulah Dethrow nursed him for three months at the Sanitarium, but, she said, he dismissed her because he thought he could not afford to pay her.

Delbert Payne was general office assistant to W. J. McDonald and saw him every working day. After Henry's funeral, the distraught brother went back to his office and carried on his ordinary work. He told Payne to live a clean life, go to church, and be honorable and upright and square in his dealings. "He said to let women alone, let wild women alone, let whiskey alone. He always advised me to go into business for myself and not work on a salary all the time." According to Payne, his boss was never heard to talk in a disjointed or disconnected way. No one was too good to work, McDonald said. He had worked all his life and could not stomach loafers who dressed up and did not work—"jelly beans" he called them. Christ was a carpenter and had shed a supreme honor on labor.

Payne was present at the signing of the disputed will, and he testified that McDonald signed his name on every page to prevent substitution, then revoked his old will.

The question arose as to how long the observatory clause had been around. Ralph Speairs, a former bank employee, testified that he had typed six wills between 1916 and 1922. Judge W. F. Moore—one of the counsel for the university—described drawing two wills, one written in manuscript by Henry McDonald. Both contained the astronomy bequest. One witness had drafted a will in 1916 that did not.

Medical Testimony[11]

The nature of McDonald's illness came under intense scrutiny, but the results were so ambiguous as to provide scant comfort to the plaintiffs. Dr. W. L. Braasch of the Mayo Clinic had diagnosed carcinoma of the prostate. Dr. L. P. McCuistion, chief of staff of the Paris Sanitarium, testified that this ailment would cause intense pain—presumably requiring mentally debilitating sedatives—but he did not agree with the diagnosis. Nonetheless, one of his colleagues had prescribed an appropriate treatment.

Two more medical men appeared as expert witnesses, including Dr. Guy F. Witt, a specialist in mental diseases. They were posed a variety of hypothetical questions, on the basis of which they said McDonald was of unsound mind. Under later questioning, it came out that the conclusion would have been totally opposite had the hypothetical questions been somewhat different. It was not medicine's finest hour.

The Trial Ends, but the Tribulations Continue

Presiding Judge Newman Phillips instructed the jury at length on the law relating to testamentary capacity. Attorney A. P. Park gave the closing argument for the plaintiffs, citing the law of descent and distribution by which the estate would have been divided among the relatives had there been no valid will. Judge Moore concluded the defense by asserting that the bequest was a sound and beneficent one, citing the value of astronomy and its applied sciences to the human race. It is likely that his remarks were based largely on the arguments supplied by Frost and other respondents to Benedict's plea for support from the astronomical world.

After a short deliberation on Saturday, 11 September 1926, the jury found Mr. McDonald of sound mind and the will valid. Congratulations poured in to university officials, but the legal wrangle was far from over. The family brought a motion for a new trial, on the grounds of suppression of evidence and the judge's refusal of the request of the plaintiffs to have put to the jury a particular definition of insane delusion. The motion was denied, but the case was then taken to the Sixth Court of Civil Appeals in Texarkana.

In such an action, no new evidence is taken. The presiding judge, Associate Justice Richard B. Levy, reviewed the evidence from the lower court. The petition for the plaintiffs again cited evidences of delusion or insanity. The university attorneys countered by citing J. D. Roberts' testimony that "Mr. McDonald had his characteristics. He did not dress up, and usually wore ordinary, but good, clothes. He did not look nor act like a rich man. He did not boast nor make a show of his wealth. He was a saving man, and did not believe in wasting money—I never knew him to indulge in any extravagance or become anything other than the W. J. McDonald I had known all my life." [7]

After a lengthy and learned opinion, the judgment of the lower court was upheld on 21 April 1927. [14] It was still not the end. The case was taken to the Texas Supreme Court on a writ of error. Again all the evidence was reviewed: the bases of testamentary capacity recounted and the conduct and words of Judge Newman Phillips examined. The presiding judge gave great weight to the medical testimony of Dr. Witt, to the probable malignancy, to several stories of aberrant behavior, to the allegations of bad family relations, and to testimony that the deceased had actually read a *Scientific American* article on communion with the dead.

Finally, on 29 February 1928, Justice Ocie Speer opened his judgment with the words, sweet in the ears of lawyers and ominous to those of litigants, "This most interesting case . . ." With the concurrence of Chief Justice C. M. Cureton, he ordered that the case be returned to the District Court of Lamar County. Two grounds were cited: McDonald was of unsound mind in 1925, and Judge Phillips had erred in his instructions.

In the original trial the contestants had requested that a specific phraseology be used in explaining testamentary capacity to the jury. The actual charge read by Judge Phillips seems indistinguishable in meaning, but the requested charge was not given. Justice Speer found that "the case turns upon the correctness of the trial court's ruling in refusing this instruction." [15]

Denouement [16]

And so back they went to Paris, Texas, to start all over again. Battle was rejoined on Monday, 29 October 1928, with a vast array of legal talent, now including Attorney General Claude Pollard and his men. The witness list had swollen to 107; their evidence was the same mixture as before, but more so. Several people had heard McDonald talk about looking into Heaven and there were arguments as to whether he meant Heaven or the heavens. Autrey Burnett now said "he was off his caboose." He was said to have contemplated suicide, and to have bewailed the fact that he was impoverished by having to pay four dollars for a room when he lodged in New Orleans for medical treatment. Two eminent mental pathologists lectured the court on the three stages of senile dementia, especially in relation to his medical state. An equal cloud of witnesses spoke of his business acumen—"the smartest man in the neighborhood"—and testified to his interest in astronomy and possession of books actually showing signs of use. They spoke of his interest in other sciences—especially botany and geology—though when it came to ants, myrmecology was made to look as suspect as astronomy.

Final summation by counsel lasted for six and a half hours for each side. Speaking for the university, Tom Beauchamp said, "Take away astronomical knowledge from man, and in six months there would be a state of confusion." On the opposite side, there were several appeals to the presumably fundamentalist religious beliefs of the members of the jury. Clyde Sweeton, in a hell-fire conclusion which took a swipe at the now President Benedict, first remarked sagely that nothing had been left in the will for the maintenance of an observatory, only construction. Then he thundered, "If the University of Texas does not believe W. J. McDonald suffered from an insane delusion about the gates of heaven, then it does not believe in the Bible and is no fit place to send boys and girls." His colleague with the Shakespearean name of Touchstone added, "Don't you see why the Lord never meant for mortal man to see into the spiritual heaven?"

Finally, anticlimax. A hung jury, with ten for the will and two against, a mistrial and remission to the docket for February 1929. The contestants moved for a change of venue, which was denied. By this time everybody's patience was exhausted. The university had spent close to $80,000 in attorneys' fees, as well as other expenses, and no doubt the contestants had made an enormous hole in the legacies they would receive if the will were upheld. The contestants had already made overtures to compromise the case on a fifty-fifty basis, but the university regents refused. Finally, the two parties agreed to settle out of court for a quarter of a million dollars to the plaintiffs.

When all the legacies and bills were paid and the agreed sum given to the relatives, about $840,000 remained. So, in the end, in 1929, the university had its money and an obligation to establish an observatory, with precious little idea as to how to set about the task.

CHAPTER 4

The Texas-Chicago Agreement

An Embarrassment of Riches in Austin

The university authorities were in no position to make final decisions, so long as they were still enmeshed in the legal tangle. Nonetheless, they quickly saw the need of gathering information toward the day when the money would be in hand. Two basic questions had to be answered: what sort of an observatory to build, and where to build it. On both questions, the University of Texas had no lack of advice, some of it extraordinarily bad.

One of the first positive steps taken by the regents, in July 1927, was to send John M. Kuehne on a fact-finding mission. For $200, the University of Texas physics professor was to visit a number of the leading astronomical observatories to collect information and advice to be used in planning the construction of the McDonald Observatory. Kuehne stopped at the Lowell Observatory (Flagstaff, Arizona), Mount Wilson (Pasadena, California), Lick Observatory (San Jose, California), and the Dominion Astrophysical Observatory (Victoria, British Columbia).

Kuehne's report was as surprising as the fact that $200 could then cover this amount of travel. The leading idea was that the money should be used for an observatory close to the Austin campus, to facilitate teaching. Ignoring their own experience with mountaintop observations, many of the astronomers thus consulted opined that the hills around Austin would provide good observing conditions, despite their mere thousand-foot altitude.

The regents toyed with the possibilities of twelve sites in and near Austin. Apart from the campus itself, they thought of rural areas—not too far out, but well clear of the city—such as Mount Barker, or Cat Mountain, or Jollyville, or Anderson Mill, or Oak Hill. Happily, no action was taken. All these are now well-developed suburbs of the expanding city.

Even assuming that the lawsuits would eventually be won, the university authorities were reluctant to press on too hard. They had in view a sum adequate to construct an observatory, but not to equip and maintain it. In those pre-inflation days, they seriously considered allowing the capital to accumulate for thirty years before undertaking construction.

With the legal battles behind them, the trustees began to deal with the estate. In late 1929 and early 1930, they received real estate, stock securities, loan notes, and cash totaling slightly more than $800,000. Lawyer Fleming was retained to liquidate the real property and promissory notes. The first expenses extraneous to the legal and fiscal matters were payments to Constance McCuistion and to Charles and Leila McDonald for information about the benefactor. This was used in the compilation of a biography of William Johnson McDonald, published in *McDonald Observatory Contributions*, volume 1, number 1. The authors were Mamie Birge Mayfield, a Fort Worth journalism teacher and librarian, and Paul M. Batchelder, professor of mathematics at the university.

In reality, the regents realized that there was yet a third problem that was far more serious than the site or the telescope. The University of Texas had no working astronomers with which to staff an observatory. It did not take long, however, for this problem to be solved in a unique way. The solution required another institution with a complementary set of problems.

Winters of Discontent in Williams Bay

Far to the north, in Williams Bay, Wisconsin, the staff of Yerkes Observatory was restive. Their largest instrument, the gargantuan 40-inch refractor built by Alvan Clark and figured by the elder C. A. R. Lundin, had magnificent optics, but it was now too small for many of the studies that had come to the forefront of astronomy. In addition, the winter weather on the banks of Wisconsin's Lake Geneva was often poor. Some of the astronomers, including Otto Struve and George van Biesbroeck, thought that an outstation should be established in another part of the country, some place with clearer skies.

There was a brief flare of hope early in 1928. A Chicagoan named O. M. Abbott had let the University of Chicago know that he would leave a bequest for astronomical research. President Max Mason asked Edwin Frost for suggestions about how such monies could be used to improve research and instruction in the Department of Astronomy and Astrophysics, of which the observatory was a part. "We need a reflector of either 60 inches or 72 inches aperture," he replied.[17] Frost would have been satisfied to have such an instrument on the Yerkes grounds, addressing one problem at a time. Van Biesbroeck built a clay model of a modified version of the Yerkes main building, with an extra wing for the new telescope dome.[18]

The cost of the new reflector was estimated at $300,000 to $400,000. If Mr. Abbott did not have that much to give, perhaps the announcement of his proposed bequest would trigger other gifts to fill out the need. Alas, the plans were shattered when an official visitor called on Mr. Abbott and discovered him living in very meager circumstances. "The possibility of making a bequest [is] probably imaginary," Frost was told. The visitor "distinctly felt it to be a psychopathic case."[19] This strange affair may, in fact, have been induced by the publicity surrounding the McDonald trials.

While Frost was willing to install a large telescope at Williams Bay, some of the younger astronomers were already chafing at the climate there. The dry (and warm) southwest had a seductive charm for them. Struve and van Biesbroeck had already investigated the weather records, and it seems a curious coincidence that their choice had fallen upon Amarillo, Texas, as potentially being a good place. It was surely for this reason that Frost was able to advise Benedict in 1926 that "the high lands of western Texas might prove to be very suitable for both solar and sidereal investigations."[20] In retrospect, it seems almost inevitable that the Yerkes and McDonald observatories would somehow be joined by more than simple collegiality.

Benedict and Struve

Frost had a great deal of influence on the Texas situation, much more than posterity has credited him with. Nonetheless, the establishment of McDonald Observatory on correct lines was due almost entirely to two men, Harry Yandell Benedict at the University of Texas and Otto Struve at Yerkes Observatory. One of them had money for an observatory, but no astronomers; the other had a respected astronomical organization in need of new facilities. The two had a common goal: to use the McDonald money to establish one of the world's leading astronomical research centers.

Benedict had studied astronomy at Harvard under Asaph Hall, the discoverer of the satellites of Mars. On taking his degree, Benedict chose teaching rather than research, and thus his personal contributions to the science are essentially nil. His merit as teacher at the University of Texas eventually brought his elevation to a deanship. His administrative acumen was so well regarded that he succeeded President Splawn in 1927. Thus, during the crucial period when the McDonald legacy was being won and when the observatory was being designed and built, the university had a chief with a first-class astronomical background and excellent administrative capacity. Benedict was able to resist the pressures brought on by the wide publicity of the will case. He could also distinguish good astronomical advice from bad. It is for that reason that there is not now a large and useless astronomical facility in the middle of the burgeoning capital city of the state of Texas.

At Chicago and Yerkes, Otto Struve had become a presence in the wings well before the McDonald bequest was adjudicated. He came of a great dynasty of astronomers. His great-grandfather, Wilhelm Struve, had been director of the Dorpat (now Tartu, Estonia) Observatory, and then founder of the Pulkovo Observatory near St. Petersburg (Leningrad). His grandfather, also Otto, had been director at Pulkovo; his Uncle Herman was director at Koenigsberg in East Prussia; his Uncle George was founder and director of the observatory at Berlin-Babelsberg; his father, Ludwig, was professor of astronomy at Kharkov. The younger Otto would live up to the family standard. He was already on the way.

This Otto Struve was born in Kharkov on 12 August 1897. He began the technical education necessary to become an astronomer under the watchful eye of his father at Kharkov University. His upper-class life was interrupted first by a tour in the Imperial Artillery during World War I, then by service in the White Army during the postrevolutionary civil war of the early 1920's. Narrowly escaping from the Crimea after the defeat of Wrangel and Denikin, he became a refugee laborer in Constantinople until his situation became known to the astronomical world through the

International Red Cross. Edwin Frost offered him a position at Yerkes which, with the financial assistance of a Chicago physician named Isham, he was able to take up in November 1921.

History and his colleagues record Otto Struve as a determined and capable man. Thus it is only mildly surprising that he was already the power behind the throne in Williams Bay long before he was director there. Frost was a good and conscientious man, but he had become blind and was getting old. As time went on, more of the business of preparing for the future of the Yerkes Observatory was either placed or let fall into the hands of the young Russian. By the time Struve actually became director, on Frost's retirement in the summer of 1932, the future was already sealed, on Struve's terms.

Contact and Negotiation

From the moment that the McDonald bequest became public knowledge, Benedict had adopted the policy that it was better to ask advice from experts than to grope wildly in uncharted territory. He did not know the astronomical issues well himself, but he did know whom to ask. He also seems to have known how to evaluate the responses. The first advice that he got was from Frost at Yerkes, and it was among the more sensible. Frost had also been extremely generous with good counsel in response to Benedict's plea for ammunition for the trial defense. Then there was this most interesting idea from Yerkes about a collaboration . . .[21]

In later years, Struve always represented the initiative for an agreement between the Universities of Texas and Chicago as his own. Certainly, for many years he had felt the limitations of a prestigious but financially strapped private university, had been irked by an indifferent climate and lack of a large reflecting telescope. Had not he and van Biesbroeck long ago proposed an outstation of Yerkes at Amarillo? Struve maintained that Texas was Struve's idea.

Maybe it was. Maybe it wasn't. The Yerkes director's archives are silent on this point, but John M. Kuehne from Austin visited Williams Bay twice during the critical period to gather more information on working observatories. One of these visits occasioned a social chat in lawn chairs on the roof of the main building, attended by crack optician Frank E. Ross and the young astronomer William W. Morgan. The problem of establishing an observatory without astronomers was being discussed when, Morgan recalls,[22,23] Ross countered that the Yerkes problem was quite opposite, a bunch of astronomers without a major telescope. Maybe there was an accommodation to be made here. The seed was planted.

What is beyond doubt is that it was Struve who conveyed the idea to Frost, who authorized Struve to follow it up and devise a plan. Once in possession of the ball, Struve carried it with fierce determination, being fully convinced of his own value as both a scientist and an administrator. From then on, he dealt directly with the presidents of the two universities.

Much of the initial probing was done by telephone, including the last informal step, a call from President Robert M. Hutchins of Chicago to President H. Y. Benedict of Texas. Hutchins proposed a formal collaboration, and Benedict accepted in principle. At Benedict's request, Struve came to Austin for a luncheon meeting with several faculty members on 14 April 1932. Hutchins' proposal was formalized in a letter dated 18 April 1932, which was presented to the regents of the University of Texas five days later. The speed with which things were suddenly moving is attested by the fact that the first contact with the Warner and Swasey Company, eventual builders of the telescope, had already been made ten weeks before.[24]

That same month, Struve was already indicating a preference for the Davis Mountains, abandoning Amarillo. The mountains, he said, "will give better promise of good observing conditions."[25]

The proposal was both simple and bold. The University of Chicago would enter a compact with the University of Texas to provide the astronomical expertise to construct and operate the McDonald Observatory in symbiosis with the Yerkes Observatory. There would be a common director, a common staff, but separate fiscal responsibilities for the two schools. Both sides thought the idea admirable, but working out the details was not exactly straightforward.

Struve worked meticulously over the draft agreement. He knew what he wanted, and every phrase should serve that want. Perhaps the most telling point is that the modifications to a vague directorship clause still exist in his own hand. The new and final version provided that the director of Yerkes Observatory (to which position he was already nominated) would be ex officio director of McDonald, and that the interuniversity agreement would automatically be subject to renegotiation upon the resignation or removal of the director. Despite the Texas position depending on that at Yerkes, the director of McDonald would report directly to the president of the University of Texas, not through the University of Chicago. Struve also insisted that the agreement stipulate that the observatory "will be built and equipped according to the plans of the Director and under his supervision."[26] Struve was going to have the best of both worlds, and he was building a few useful levers into the control mechanism.

Struve's precautions were only the most overt evidences of a general fear among the Chicagoans. What if the Texans are clever enough to arrange that Yerkes staff and cash help them put a major observatory into working order, then throw them out before they can get any benefit from the effort? The proposed clause that either party could terminate on two years' notice was particularly worrisome. "If," noted Struve, "Texas should give us notice one year before the telescope is completed, all of this would be a total loss to us."[27] Despite the Chicagoans' worries, this clause was eventually strengthened in Austin's favor, reducing the notice period to only six months.

Compact

Eventually, the barriers of mutual distrust and fear were broken down, and agreement was reached. The Texas regents accepted Struve's requirements for the directorship with no change. The two universities agreed to a thirty-year collaboration in which the University of Chicago would operate the McDonald Observatory for the University of Texas. Each separate university accepted specified responsibilities for the staffing and operation of the observatory.

The agreement required the University of Texas to build and equip the McDonald Observatory in the state of Texas, at a cost not to exceed $375,000. The plans were to be prepared by the director, subject to regental approval, with a completion date of 1 July 1938. Struve had obviously done a great deal of homework; appendix Exhibit A provided for the following:[28]

Mechanical parts of telescope	$70,000
Revolving dome	70,000
Optical parts (mirror of approximately 80 inches in diameter and focal ratio 1:4, Newtonian and coudé flats, Cassegrain mirror)	80,000
Buildings and site	60,000
Installation	20,000
Site selection	5,000
Accessories (coudé spectrograph, convertible Cassegrain spectrograph of glass, quartz spectrograph, photometer, plateholder, microphotometer, corrector, two Ross wide-angle lenses)	45,000
Carpenters' and technicians' shop equipment	3,000
Truck or car	2,000
Margin for power line, telephone line, road work, etc.	10,000
TOTAL	$365,000

Already, the putative director had an almost complete design for an isolated observatory in his head, even though the financial allocations now strike us as pitifully small for the construction of the second largest telescope in the world.

The remaining obligations of the regents were to appropriate $10,000 per annum toward running the observatory, to produce "Publications of the McDonald Observatory of the University of Texas" (starting with a biography of the benefactor), to promote the library, to pay for observational tests of possible sites, and to allocate time to personnel from Austin, Yerkes, and possibly elsewhere, for the use of the telescope.

Chicago, for its part, would appoint an assistant director to live on the site, together with two observing assistants, an engineer, and a janitor, and pay their salaries. The salary total was expected to be $23,000 per year. Chicago also agreed to provide a site tester and sundry expenses for photographic materials, transportation, and maintenance, bringing the total cost of running the observatory to about $40,000 per year.

The agreement would continue in force for thirty years, subject to cancellation by either party on six months' notice, or on inability of the parties to agree on the appointment of a successor in the event of death, disability, or resignation of any director.

Upon their ratification of this compact on 23 November 1932, the Texas Board of Regents then viewed models and drawings of the proposed observatory, with the best bid already submitted by the Warner and Swasey Company, of Cleveland, Ohio.

The World Reacts

Today such an agreement might not seem exceptional, but at that time, before the days of big science, such a contract between a celebrated private university and a relatively obscure southern state university, each jealous of its autonomy, was little short of revolutionary.

The negotiations, while delicate, were far from secret. It was Struve himself who saw to that. With approval from the Chicago administration, he sent letters to Henry Norris Russell (Princeton Observatory), Frank Schlesinger (Yale Observatory), Harlow Shapley (Harvard Observatory), and George Ellery Hale (Mount Wilson Observatory) to explain the Texas-Chicago plan while the details were still being worked out, and to ask their advice.[29] Struve's argument for doing so was that the plans were so far advanced that confidentiality was no longer necessary, and that those inclined to make difficulties "will do so without our telling them about it."[30]

Shapley and Schlesinger Worry about Yerkes

Struve's dean had opined that it would be a mistake not to consult Shapley, that Schlesinger would be helpful, and Russell would too, "as soon as he recovers from his grouch."[31] The estimate was squarely on the mark. Russell caused no trouble. Both Shapley and Schlesinger were generally supportive, but with some very trenchant worries about the disadvantages for Yerkes.

In his response, Schlesinger provides fuel for the doubt that Struve was the original author of the idea. "I realize how difficult it would have been . . . for you to have turned down this proposition after it . . . had been made to your President." Still, his primary concern was institutional. He predicted that Chicago would now spend money on McDonald that it should properly be spending in Williams Bay, that these expenditures would eventually exceed the McDonald bequest, and that large telescopes were not really essential. Nonetheless, the decision made, he was behind it. "It will be great to watch your activities."[32]

Shapley's fears were quite different. They were political and personal. Politically, he feared the entanglements of state institutions. "Perhaps the Regents are pretty independent of politics (That is one of the wildest hypotheses that I have ever made) or perhaps Pres. Benedict . . . can handle the situation." He, like Schlesinger, was not convinced that a giant telescope was necessary for astronomical progress, but he did recognize its potential as an expression of the Texan personality. He swung behind it with "if you stay with the plans, I have no doubt that [the telescope] will be completely justified." He was not convinced, however, that the personal sacrifice that would be required of Struve—in terms of his own research program as much as of his energy—was worth the effort. As for the dissentient voices, "They will all come around."[33]

See Worries about Texas

Yerkes was not the only beneficiary of champions who worried for its good. Captain Thomas Jefferson Jackson See, U.S. Naval Observatory, ranked in his own eyes as the foremost astronomer of the age. To almost everyone else, he was a once-licensed but now-suspended buffoon who had been barred from the pages of the *Astronomical Journal* for his vituperative excesses. He had met Benedict during the latter's student days and used this excuse to send him a fifty-page memorandum on the efficiency of American observatories.

See resembled the Marx Brothers' character who sang "Whatever it is, I'm against it." He was against astrophysics, the *Astrophysical Journal,* external galaxies, reflecting telescopes, unnecessarily ostentatious buildings (as at Yerkes), and Edward C. Pickering at Harvard and George Ellery Hale at Mount Wilson (they hung around with millionaires and sought publicity). Most observatories were inefficient—Harvard and Mount Wilson only rated 12 and 6 respectively out of 100, and the average was only 46. McDonald Observatory ought not to have permanent buildings; a wooden library would do. There was an 18-inch lens to be picked up cheaply, and maybe the transit circle and zenith tube could be backed up by a fair-sized refractor. See owned a farm near Tyler that they could buy (Tyler is in the low, marshy woodlands of Northeast Texas). Finally, with underlinings worthy of a polemicist, he wrote, "the *Bungling* and *Unworkable plans* you have formed with the Yerkes Observatory: *it will not stand five years!* It [sic] carries with it *divided responsibility* and *divided credit for results, if any are attained, by the foreigner entrusted with matters which should be in the hands of a real American!*"[34]

Struve himself reported that the classic response was supplied by an unnamed member of the Board of Regents of the University of Texas: "Better a foreigner than a damned Yankee."[35]

The *Ciel et Terre* Affair

In the archives of the Yerkes Observatory today, one and only one issue[36] is missing from the 1934 volume of the semitechnical astronomical publication *Ciel et Terre,* or indeed of any nearby year. Thereby hangs a curious tale of tactless indiscretion. That issue contains what is described as excerpts of private correspondence between the van Biesbroecks and a friend. One letter evidently written by Mrs. van Biesbroeck tactlessly describes the role of her husband and Struve as the two most important people and makes it sound as though "the two Europeans" were going to use the Texas money to set up an observatory solely for their own benefit. It seems clear from subsequent correspondence that Struve thought that the issue never arrived simply because the editor of the Belgian magazine, E. Lagrange, did not have the courage to send it to Williams Bay.

Struve's attention was called to the matter by Harvard's Bart Bok, and he reacted angrily. In Struve's eyes, publication of the letter was very compromising, both because it misrepresented the very delicate relations with Texas and because it was not a good era to flaunt European origins in the American heartland. Benedict's experience with See is evidence of that. Both Bok and his director, Shapley, counseled restraint, as they considered the letter more an embarrassment to the author and the journal than to Struve and Yerkes. In fact, Shapley identified the problem as a part of the rivalry between California and "eastern" institutions, of which we will encounter more later. Lagrange responded to Struve's choler with a wishy-washy letter, and the issue was forgotten.

Relations were not going to be entirely cordial with the West Coast for much of two decades, but Struve plunged ahead. Optician Lundin Junior (then of C. A. R. Lundin, Inc., Astronomical Telescopes Visual and Photographic, Watertown, Mass.) congratulated Struve on his accession to the Yerkes directorship with a photograph of Struve's grandfather and a document of his father's, found in the Alvan Clark archives. The papers were to be delivered by Edwin P. Burrell, of Warner and Swasey.[37] Separately, Benedict asked Struve for a professional evaluation of a used 9-inch refractor that Warner and Swasey had offered in response to a Texas request for bids on an instructional telescope to be mounted on the new Physics Building. On the basis of Struve's report, the refractor was purchased, and it is still on the Austin campus. The relations for the telescope contract were being cemented.

As for the astronomers, Shapley's prediction came true in large measure, at least in the East. Presented with an accomplished fact in the form of a Texas-Chicago compact, they were generally supportive, but in fact Struve did not have much need of them. What he did need was local help in West Texas to secure the site he wanted, continued support from the Texas Board of Regents, and competent craftsmanship to build the world's second largest telescope. And patience.

CHAPTER 5

Choosing and Developing the Site

The most immediate task was to find a place to put the McDonald telescope. There were plenty of suggestions. In the five years following publication of the will, the university had received thirty offers or proposals of land "suitable for an observatory."

The will did not specify that the observatory had to be in Texas, although it clearly carried some implication of this. Mount Wilson's Edwin Hubble wrote to suggest a southern hemisphere site; indeed, in that event he just might be available for the directorship. The course of astronomy might have been far different had Hubble's proposal been realized.

West Texas figured prominently in the early suggestions. The Del Norte Mountains near Alpine and the Franklin Mountains outside El Paso were proposed early in 1926. The following year, Marfa's J. C. Fuller nominated the second-highest peak in Texas, 8,378-foot Mount Livermore in the Davis Mountains.

Back in Austin, a common reaction was that El Paso might be suitable, because of its easy access. Fort Davis and Alpine, while closer to Austin, were far too remote from transport facilities. A week-long visit by the eminent Dutch astronomer Willem de Sitter in December 1931 was partially devoted to site discussions, but these ended inconclusively. In any event, no one was willing to make a decision just then. Someone would have to lead them into it. That someone appeared on the scene when President Benedict received a telegram from Chicago on 4 April 1932. Would it be convenient, the message from Robert Hutchins inquired, for Benedict to see "the Director-elect of the Yerkes Observatory"?

Where Do You Put a Telescope?

The Texas-Chicago agreement focused attention once again on West Texas. Even before formal contact had been made, Struve

used the mid-twenties studies of the Texas Panhandle to convince
his dean and president that the climate there was good for astronomy. A fortnight after his first trip to Austin, Struve had reset his
sails: "I believe that the Davis Mountains will give better promise
of good observing conditions."[38]

Belief is not proof. Measurements would provide that. A field
expedition was prepared under the leadership of Yerkes astronomer Christian T. Elvey. Before starting, Elvey did his homework
and produced a thorough report on the weather, geology, and
topography of Texas. Armed with this, he set out on his astronomical odyssey on 13 June 1932, picking up Amherst astronomer T. G. Mehlen in Des Moines to serve as assistant. They
crossed and recrossed the state of Texas in their little Chevrolet
van, taking measurements.

The observing scheme was designed to rate each site for atmospheric stability (called *seeing* by astronomers), image quality,
and atmospheric transparency. For the first two, the pair carried
a telescope of 4-inch aperture, 3-foot focal length, and a com-
pound microscope eyepiece of 756× magnification. To simplify
the data reduction, its only target was the pole star, Polaris,
which is always seen through a constant air thickness.

Seeing was measured by comparing the average oscillation of
the image of the star to the diameter of its diffraction pattern.
This gives a reliable measure, because the diffraction pattern is of
constant size for a given instrument. The observations were recorded on a scale running from 1 (bad) to 7 (excellent), corresponding to image oscillations of from 2.0 to 0.2 diameters of
the diffraction pattern. Adjacent integer quality numbers represented factors of approximately 1.4 in the range of oscillation,
and thus were roughly proportional to the area of the seeing disk.
The seeing figures were modified to account for the effect of poor
image quality.

Image quality was rated according to the percentage of the time
that the diffraction pattern was visible and whether it was sharp
or fuzzy. The scheme shown in the table was used to adjust the
seeing score.

The adjustment was subtracted from the raw seeing figure to
give the final score. A VPP image produced either 0 or 1 as a final
result, while NP meant that the image was so bad that the test
could not be made.

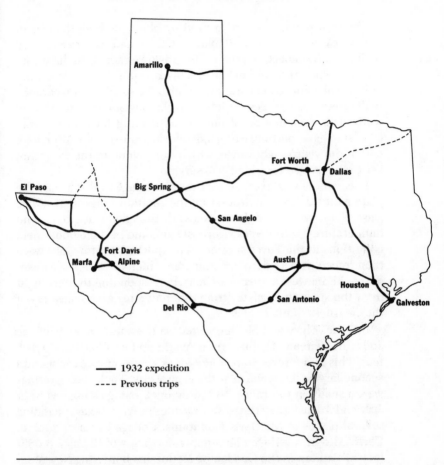

Journeys of Elvey and Mehlen in search of the perfect observatory site. Re-drawn from the original document in Barker Texas History Center, University of Texas at Austin.

The transparency was measured by photographing the bright, blue star Vega on Eastman 40 plates, with a small camera equipped with a quartz objective prism. The prism refracted the light into the star's spectrum, permitting the investigators to determine not only the absolute transparency, but also the relative transparency in different color ranges, which is very important in astrophysical work. The combination of photographic emulsion and the specific star gave particularly good information on the ultraviolet end of the visible light range, which was pivotal to the work that Struve envisaged for the new observatory.

The camera was fixed, so that the star image trailed across the plate as it was carried around by the rotation of the Earth. The glass plates were then developed with the chemicals, time, and temperature controlled as rigorously as could be done under field conditions in the Texas summer. The finished photographs were then traced with a microdensitometer, which measures the opacity of the images, a greater opacity corresponding to more light from the star. Thus, plates from different sites gave measures of the clearness of their skies.

In June, Elvey and Mehlen tested on the southern edge of the Jollyville Plateau, 15 miles from Austin and at a height of 1,100 feet. This laid forever to rest any ideas about putting the world's second largest telescope near the capital city. Their seeing ratings were usually in the range 0–2, only once rising as high as 5. In July and August, they tested several sites in West Texas, including several places in the Davis Mountains, College Hill near Alpine, Cerro Alto in the Hueco Mountains northeast of El Paso, 8,085-foot El Capitan in the Guadalupe Mountains bordering New Mexico, and Mount Franklin north of El Paso. A side trip to Flagstaff and Pasadena permitted the siteseekers to calibrate their setups against measurements made with the Lowell Observatory 24-inch refractor and the Mount Wilson 100-inch reflector.

Strengthened by this confirmation of their methods, Elvey and Mehlen returned to Williams Bay enthusiastic about the Davis Mountains. Their final choice was Black Mountain (now Spring Mountain), on the eastern edge of the Davis Mountain Range with an elevation of 7,550 feet. The runners-up were nearby Blue Mountain, on the southern edge of the range, and Little Flat Top in the heart of the Davis Mountains. The latter was the lowest of the three, but the observations on five nights there had been "unbelievably good," especially in the ultraviolet.

Image	Meaning the diffraction pattern is	Adjustment
GP	Visible all the time	0
FP	Visible ¾ time	1
PP	Visible ½ time	2
VPP	Visible ¼ time	Variable
NP	No pattern visible	

The pair of prospectors returned home after having covered 11,000 miles in the van and nearly 2,000 more in other vehicles. Subtracting the roundtrip distance from Williams Bay to Pasadena via Flagstaff, this means that they traversed nearly 8,000 miles within the state of Texas, roughly one-third of the distance around the Earth. The trip was so short that important seasonal variations would not be taken into account. Indeed, long-term records show that in West Texas the summer skies are slightly worse than the average there, but in searching for an observatory site that may be considered an advantage as easily as a disadvantage. It is as important to have the best "worst" as to have the best "best." In any event, both shortcomings would be rectified by subsequent forays.

Struve came to verify the results himself, camping out on several of the summits in the vicinity with his wife, Mary. A rapport with the townsfolk in Fort Davis had already been established by Elvey and Mehlen, and this quickly carried over to the entire Yerkes staff. One of the central characters in this was Walter S. Miller, who ran the Limpia Hotel, where many of the astronomers were to stay in the early years. Another was Barry Scobee, a tenderfoot from Missouri who had come to Fort Davis to follow a life as writer, newspaper stringer, and eventually Justice of the Peace. One night while the Struves were camped on Spring Mountain, Miller and Scobee decided to visit. When they arrived on top of the mountain, Mary Struve invited them to stay for a campfire dinner. Alas, she showed that she was even more of a tenderfoot than Scobee by putting a can of beans into the fire unopened. The inevitable explosion showered them all with half-warmed beans, shortly followed by a shower of apologies from

Mary Struve. No damage was done, and tentative friendships were cemented firmly.

Struve rejected Spring Mountain after several nights on its summit because, although it offered marvelous views, it possessed only a limited rocky area. Blue Mountain, on the edge of the plain to the south, "seemed to us to expose to a dangerous discontinuity of weather conditions." Little Flat Top, also called Mount Fowlkes, was not bad. It was not so attractive as its bigger neighbor, variously called Flat Top, U Up and Down Mountain, and Mount Locke. That is the mountain that Otto Struve wanted for the McDonald Observatory.[39]

The competition was not quite ready to give up. In December, Dr. H. W. Morelock, president of Alpine's Sul Ross State Teachers College (now Sul Ross State University), reported to Struve that local funds were available to build a road up Mount Ord "or some other suitable location in Brewster County for the observatory." When the astronomer insisted that Mount Locke was the final choice, Morelock accepted the news with perfect grace and offered his services to facilitate the establishment of the McDonald Observatory on Mount Locke in any way that he could. He had already undertaken to lobby a power company to supply electricity to the mountain, and Struve immediately asked President Benedict to appoint Morelock as the University of Texas' agent to deal with the citizenry of the surrounding area, especially the Chambers of Commerce in all the nearby counties. As a mark of gratitude for this gesture, the Sul Ross trustees assigned an office and a classroom to the Yerkes/McDonald staff, and a close relationship was maintained for many years.

A Friendly Condemnation

Mount Locke is 6,809 feet above sea level, 10 miles northwest of Fort Davis, and 16 miles along State Highway 17, the Davis Mountains Scenic Loop. Its geographical coordinates are 104 degrees 01.3 minutes west longitude, 30 degrees 40 minutes north latitude. "U Up and Down" is the name of the ranch on which the summit is located, and from which it took one of its informal names. The name describes the cattle brand, which can be simulated by juxtaposing two horseshoes, the left one with points up, the right with points down. U up and down. Local legend has it that the founder, physician G. S. Locke of Concord, New Hampshire, won the first tract of the ranch in a poker game on a coast-to-coast train ride. The land was now held by his granddaughter, Mrs. Violet Locke McIvor. It was necessary to persuade her to give it up.

The prospective location of a large observatory in their midst caught the enthusiasm of the local populace very quickly. Many of the citizenry offered their services in whatever capacity could be useful, sometimes acting without even asking authorization from the astronomers or the universities. W. S. Miller was particularly keen to have a Davis Mountain site, as he had been also a major promoter of the Davis Mountains Scenic Loop, a state highway segment through the mountains north and west from Fort Davis. According to his younger son, Keesey, the hotelman "spent a few thousand dollars, untold amounts of time, and wore out two automobiles" on these two projects.[40] He bent himself to the problem of the observatory's need for Mrs. McIvor's land. Struve and Benedict had the good sense to appoint him as one of their local representatives for the negotiations. One of his more overt moves was to propose that the name *Mount Locke* be submitted formally to the U.S. Commission on Geographical Place Names. After all, he noted, there are several peaks in the area called Flat Top. Mrs. McIvor was greatly pleased at this gracious remembrance of her grandfather.

The first move was a straightforward suggestion to Mrs. McIvor that she donate the 200 acres around the peak to the university for an observatory. There were problems. Mrs. McIvor was sympathetic, but she was only the trustee of the property, and she had already given 200 acres from that same section for a state highway park. She suggested that the state secure another 200 acres and trade it for the observatory site. Commenting on this proposal, W. W. Negley of San Antonio advised Benedict, "Mrs. McIvor is a woman of integrity and strong character. When she makes up her mind, it usually stays made up."[41] Another San Antonian, Robert L. Holliday, sought the aid of Fort Davis Judge Edwin H. Fowlkes (owner of Little Flat Top) in arranging the proposed swap.

The swap idea seemed to carry the day; the Union Trading Company (of which W. S. Miller was a director) came up with the necessary collateral. Van Biesbroeck, in Fort Davis on a reconnaissance mission, was able to report that the deed to Mount

Locke was received by mail from New Hampshire on 17 April 1933. It was recorded by Jeff Davis County Clerk Herbert D. Bloys, but only after he had telephoned H. Y. Benedict in Austin. Within twenty-four hours, the state legislature received a bill authorizing the construction of a road to the top of Mount Locke. It passed.

In a flash of prescience, Miller suggested to Struve that it might be good to have the smaller peak as well as the larger, "for future development, in case the Rockefeller Foundation should join in with you some day." [42] It wasn't a bad idea.

Judge Fowlkes had died during this period, but a second parcel of 200 acres around Mount Fowlkes was secured with the cooperation of Mr. Bloys and the administrator of the estate, Nicolas Mersfelder. The title deed to the Fowlkes donation was dated 3 August 1933, and contained the provision that the land should revert to the trust if it were ever used for any purpose other than an observatory.

Struve had his mountain, but not for long. One of the regents, Judge Robert Lynn Batts, became concerned about whether Mrs. McIvor could legally disperse parts of a trust in this way. If not, there might eventually be a challenge to the university's title to the land. This possibility created a great deal of apprehension, especially in view of unfounded rumors in Fort Davis that other towns had offered large sums to have the observatory built elsewhere. Warner and Swasey asked for and got a contractual commitment from the university for a reimbursement if "by reason of uncertainties of title . . . a change in said site is made by the Trustees." [43]

Legal problems have legal solutions, especially if you are the state. Someone suggested that the parcel be named in a friendly condemnation suit, in which the state of Texas exercised its right of eminent domain to the satisfaction of all parties. The attorney general's office filed the action in June 1935. This maneuver assured that the title was rock solid for all time.

Site Preparation Is a Community Affair

Van Biesbroeck reported the receipt of the McIvor quit claim deed when on the first of two expeditions to complete the site testing at Mount Locke. Struve sent him off in April 1933 to test seeing, survey the mountain, determine the meridian and the geographical coordinates, and inspect other possible sites "in case Flattop should not be available."[44] Although he went down to sleep at the hotel by day, at night the tiny Belgian camped out atop the mountain, marking a trail up from the highway by tying bits of cloth or string on the bushes.[40]

What he found on his arrival was an astonishing flurry of local initiative. W. S. Miller had, in fact, got the state parks commissioner to release the McIvor land in favor of the observatory before the formal deed had arrived. Anticipating the desires of the University of Texas, he had talked the Jeff Davis County commissioners into authorizing a preliminary survey of the roadway to the top of Mount Locke, while rancher J. W. Merrill had led another group to survey the boundaries of the two tracts that would eventually comprise the observatory property. A line had even been laid out from a nearby spring to the foot of the mountain. With a touch of wonder, the astronomer remarked, "They seem to take it for granted that their doings will soon be okayed."[45] Their faith was justified, at least in part. The road that they laid out was built later that year by the State Highway Department.

The developing social bond between the townsfolk and astronomers continued to progress during Van Biesbroeck's stay, when the two Miller boys, Espy and Keesey, and their wives, Lucy and "Bit" (whose real name, Lillian, is unknown to most of her close friends), climbed Mount Locke one night to look through the telescope.

There was no obvious source of water near Mount Locke. Indeed the first official contact between Mersfelder and the regents was an abortive proposal concerning water rights and a pipeline. To pipe water from Spring Mountain would have been far too expensive. So would pumping water from Limpia Canyon—although that is what is now done. Nonetheless, the McDonald trustees had made a commitment to provide a water supply by 15 September 1933. State geologist, Dr. Hal G. Bybee chose a drill site on the level ground to the northwest of Mount Locke. When President Benedict and the regents first visited the mountain in August, rancher Joe W. Espy treated the party to a chuckwagon dinner at the proposed well site.

Bybee and Mr. A. M. Barnes of Fort Stockton began drilling at the end of August 1933. Some 400 feet down, the drill entered a pocket that kept collapsing, and no progress could be made. Bybee decided that they needed 10 quarts of nitroglycerine, which could be obtained from the village of Wink some 140 road miles to the northeast. Barnes pointed out that it would cost $65 to have it brought, but only $20 if it were fetched. Bybee did have a car. Bybee was not overjoyed by the thought of driving that far over indifferent roads with so much touchy explosive on board. In a burst of generosity, he offered the car to Barnes. The driller returned safe and sound the next day with the juice and a bill of $1.37 for seven gallons of gasoline.

The pocket vanquished, the drillers struck water at 818 feet and first pumped on 20 December. This well supplied the needs of the observatory for some forty years before running dry.

By the time the well was finished, the University of Texas had signed a contract with the Warner and Swasey Company for the construction of the McDonald Observatory at Mount Locke. Local initiative phased out, and big industry came in to finish the job.

CHAPTER 6

A Mosque on Mount Locke

During his April 1933 expedition, George van Biesbroeck had essentially platted the mountaintop. In addition to fixing the true north and south directions, he chose the location for the telescope dome and noted possible sites for the principal residences. During the construction of the observatory, he would often be on the scene, drawing on his training as an engineer to assure that Struve's instructions were carried out.

Even in 1933, $800,000 was not an enormous amount of money with which to build and equip an observatory. The University of Texas had to find ways to minimize costs. It was decided that one way was to find a single reliable project manager to build everything: the telescope, the optics, and the observatory. This had never been done before with a major instrument. Indeed, there had been only one previous major telescope built entirely by American firms, even with divided responsibility.

The possibilities were limited, and in practice reduced to two: the J. S. Fecker Company of Pittsburgh, Pennsylvania, and the Warner and Swasey Company of Cleveland, Ohio. Struve had experience of major telescopes built by both firms. Fecker had built the fork-mounted 61-inch reflector at Harvard. The 72-inch reflector for the Dominion Astrophysical Observatory at Victoria, British Columbia, and the 69-inch reflector at the Perkins Observatory in Delaware, Ohio, had both been built by Warner and Swasey. Up to this time, however, each firm had been responsible only for the mechanical parts of the telescopes.

A Package Deal from Warner and Swasey

During a boat excursion at the 1932 General Assembly of the International Astronomical Union at Cambridge, Massachusetts, Struve encountered Edwin P. Burrell, director of engineering at Warner and Swasey. The company, Burrell told Struve, had started an optical shop with the younger C. A. Robert Lundin in charge. The Warner and Swasey Company was prepared to bid on the construction of the entire observatory, from mechanism to optics to building.

The Warner and Swasey Company occupied a unique place in American astronomy. The company had been founded in Chicago in 1880 by two young New Englanders, Worcester R. Warner and Ambrose Swasey. They didn't like Chicago and moved to Cleveland the following year. Their purpose was to build machine tools, and they rapidly acquired a reputation for fine precision products. Both of the founders were deeply interested in astronomy, and they used their shop to build a 9.5-inch equatorial telescope during their first year together. After that, fabricating telescopes became a passionate, though not profitable, sideline for the company. In addition to the two large reflectors cited above, their products included the mechanical parts of the Lick Observatory 36-inch refractor, the Naval Observatory 26-inch refractor, and the Yerkes Observatory 40-inch refractor. Mr. Swasey, now retired but still quite active, wanted very badly to build the world's second largest telescope for the University of Texas. He got his wish and lived long enough to see the movable dome structure assembled in Cleveland before it was shipped to Texas.

The contract for the telescope was signed by the trustees of the McDonald Telescope Fund, a committee of the whole of the University of Texas Board of Regents, on 19 October 1933. It was ratified by the Warner and Swasey Company on 27 October. It stipulated that the site had been chosen, that Struve was the observatory director, and that Warner and Swasey would build in consultation with him according to comprehensive plans and specifications agreed among him, Warner and Swasey Vice-President Charles J. Stilwell, and H. Y. Benedict for the University of Texas.

On the mountain, Warner and Swasey were to be responsible for the construction of the dome to house the telescope, but under the watchful supervision of the Office of the University Architect. The trustees were committed, in addition to the access road and the water supply already under construction, to provide a power plant and a telephone line by early April 1934.

The university did not wait long to exercise its oversight power. An earthquake had occurred less than three years earlier, on

16 August 1931, with epicenter near the town of Valentine, about 20 miles from Mount Locke. The Richter magnitude is now estimated at 5.9,[46] which is far from trivial. It did no serious damage, but architectural engineer W. W. Dornberger insisted on extra bracing of the dome building beyond what was proposed by Warner and Swasey.

The machine-tool builders had contracted to be responsible for, not necessarily to do, everything. The steel frame and dome were subcontracted out to the Patterson-Leitch Company of Cleveland, while the on-site excavation, concrete work for foundations and piers, and final assembly of the building were subcontracted out to several "local" companies from San Antonio and El Paso. All of these firms had need of local labor, which was a welcome relief to the economy of Fort Davis during those bitter days in the deepest part of the Great Depression. The work rolls soon included family names such as Dutchover, Granado, and Hartnett— names still familiar on Mount Locke today.

The Telescope Building

The plans called for a steel building with the two lower floors containing a library, offices, darkroom, and bachelor living quarters. The third floor was to house the telescope and observing area, capped by a steel dome turning on rollers. This concentration of everything in a single building is far from unique (Yerkes Observatory being a prime example), but placing the observing floor for a giant telescope above the working and living areas is uncommon. It was adopted at McDonald as an economy measure.

Following the common astronomical practice, the dome was to have a vertical aperture to be closed by shutters moving in from either side. This permits the telescope to be pointed in any direction around the horizon and at any angle above the horizon, while opening only a small part of the roof structure. The smallest, and thus cheapest, dome diameter consistent with allowing an uninterrupted sweep for the telescope motion in all directions was 62 feet, and this essentially determined the size of the cylindrical building beneath. The telescope is mounted on two massive concrete piers, the northern one higher than the southern to give the correct tilt to the polar axis of the telescope. To ensure the telescope's isolation from the vibrations generated in other parts of the building, especially by the dome rotation, it was nec-

essary that these piers be founded deep in bedrock and joined by
a massive transom at observing floor level, completely separate
from the rest of the building. Heavy-duty battleship plate was
chosen as the floor surface on the observing level. Under the telescope, two semicircular sections of battleship-plate flooring were
to be mounted on hydraulic hoists, rather like automobile servic-
ing racks, to permit access to the Cassegrain focus when the telescope was not near floor level.

The third-floor location of the telescope prompted the designers to add an optional "balcony," estimated cost $1,500. This is a
steel walkway around the outside of the dome, accessible from
the telescope room. This "catwalk" (as it is now called) has turned
out to have great practical use as a vantage from which an observer may scan the sky, as a lodgment for outside equipment
such as meteorological instruments and radio antennae, and as a
means of external access to the dome and its moving shutters. It
also provides a spectacular view of the countryside in daytime or
full moonlight.

New York architect William Gehron was hired to transform
Struve's requirements into a full building design. Gehron had a
working scale model of the structure built, a standard practice intended to insure that everything fits together as intended. After
the model had served its primary purpose, the Warner and Swasey
Company sent it on a coast-to-coast exhibition tour; it now re-
sides in the Cleveland Museum of Natural History.

Furbishing the Mountain

Stilwell arrived on Mount Locke on 13 November 1933, together
with C. J. Patterson, president of Patterson-Leitch. Ground was
broken that same day, but the cementing work was delayed by the
absence of the building inspector from the university, Hugh Yantis. He showed up a week later, but building an observatory was
not the first thing on his mind, as he got out of the university automobile carrying a steel bow and fifty arrows. Van Biesbroeck reported to Struve, "Killing a deer seems to be the main preoccupa-
tion of all our University men."[47] The astronomer apparently
threw a tantrum and threatened to leave the mountain within the
hour unless the unnecessary delays ceased. They did, and in the
end Yantis did his job very carefully and conscientiously, causing
pier holes to be excavated, insisting on greater depth and strength

for the foundations and a strengthening ring around the entire foundation raft.

The sand and gravel for the concrete work came free of cost from the neighboring J. W. Merrill and Son Ranch. Some twenty men were put to work excavating it from creek beds, washing and mixing it. Yantis got very choosy about its quality, causing some concern in Cleveland about possible cost overruns. He was eventually satisfied, and the concrete was poured into the 62-foot-diameter foundations and the 45-foot-high steel and concrete piers.

Philip E. Bliss, president of the Warner and Swasey Company, reported to the regents in June 1934 that the dome would be erected in Cleveland within a few days and would then be dismantled and shipped to West Texas. The following month saw the arrival of the first construction foremen, steelworkers, and 45 tons of structural steel. The first steel was raised on 31 July. The main frame was finished by 26 September. The shutters for the dome opening, four parts each weighing 4 tons, were in place by late October. Altogether 250 tons of structural steel went into the dome's skeleton, with 50 tons of castings such as wheels and machine parts, and 25 tons of sheet metal for the sheathing. Joints were not only either riveted or bolted, but welded for added solidity.

Scobee gives a vivid picture of the site: "Tents were strewn all over Mount Locke, on any level place in the brush or in the open, wherever space could be found, with the smell of cooking rising at all hours. Many of the men had their wives with them. It was [in] the days of the Great Depression, and some of the men were on their first jobs in two or three years. Nobody tried to put on any style. They were careful of their new curiosities—cents and dollars. Some men laughed like kids and kissed their first checks. One steel worker showed his first check to this writer with a grin that was all but drowned in tears."[48] Two of them were a half-breed Cherokee named Arch Garner and a Fort Davis Mexican-American named Tommy Hartnett. Both of them would later join the McDonald Observatory staff and play important roles in this history.

The 75-foot-high steel dome was primed and coated with aluminum paint late in October. On the last day of the month, an electrical hookup from the 10 kw gasoline-powered DC generator permitted the dome to be turned, with a group of Fort Davis citizens and company representatives taking a Halloween ride. The site was essentially cleared by the end of November 1934. In all the work, there had been no serious accident.

The construction attracted reporters from far and wide, and many news stories informed readers nationwide of the progress of the new observatory. In a flight of hyperbole that seems remarkably uncharacteristic of its staid reputation, an article in the *Christian Science Monitor* claimed that the dome was "lacking only minarets to complete its mosque-like appearance."[49]

The first dwelling, designated House C, was built in 1934. By 1935, the regents were ready to consider bids for additional buildings and residences on the mountain. Five firms tendered bids in the range from $64,000 to $80,000 for building residences A (the director's house), B (originally the resident astronomer's house, now the superintendent's house), and E, as well as a garage, power plant, and water tank. Electrical work was included, but the heating system in House A required a separate bid. This was too high, so new bids were accepted excluding houses A and E. Bids for the remaining houses and some other work were examined the following year. There was evidently anxiety about the total funds available, for in the meantime an application had been filed with the State Director for Texas of the federal Public Works Administration for a grant-in-aid. This was later withdrawn, with the report that the project had been completed without the necessity of financial aid. One way or another, the former U Up and Down Mountain was becoming a habitable and inhabited place.

CHAPTER 7

Astronomy Comes to West Texas

The mountain had been used for astronomical observations for a long while before it was habitable. It had been visited, of course, on the Elvey-Mehlen prospecting tour and by Struve on his followup to that event. Van Biesbroeck's expedition in April 1933 had been at least partly devoted to testing the sky conditions. Despite indifferent seeing, he found the transparency excellent, excepting the times when dust was blowing down from the ecological disaster farther north, the Dustbowl. Under the impression that it was the worst season of the year, he remarked, "How lucky we would be [at Yerkes] with some of this bad weather!"[50]

The November expedition was to keep an eye on the construction, and van Biesbroeck greatly regretted not having a telescope available. "I got acquainted with Achernar, which shines brightly over Blue Mountain for a couple of hours in the evening. The declination is −59 degrees."[51] That meant that it was virtually on the southern horizon, and not visible from stations farther north. After only one cloudy night in a week, and fine seeing part of the time, he declared Mount Locke to be an astronomical paradise.

Struve's Early Plans

In 1933, Struve published an article in *Popular Astronomy* on his scientific plans for both Yerkes and McDonald observatories.[52] In his eyes, they were a single scientific organization.[53] In retrospect, Struve's article seems remarkably conservative. In view of his immense contributions to astrophysics later on, it is surprising that he did not now heavily emphasize this aspect of astronomy. Astrometric work included visual observations of double stars, determinations of their mass ratios, and positions of comets and asteroids, all by van Biesbroeck, who was also computing comet orbits. G. W. Moffitt would determine stellar parallaxes and accurate proper motions of nearby stars. In astrophysics, there were to be studies of the spectral classification of stars by Struve, Elvey, W. W. Morgan, and Allen Hynek, with special attention to spectral classes A and B. Struve and Morgan would continue observing spectroscopic binaries. Struve, Elvey, and Christine Westgate would study stellar rotation. Struve and Morgan would investigate stars with variable spectra, and a program of stellar spectroscopy in the red and infrared regions in cooperation with the Perkins Observatory would be undertaken by Franklin Roach, Struve, Morgan, and Ralph N. Van Arnam.

The Yerkes 40-inch refractor was the instrument of choice for direct solar photography, this to be backed up by spectroheliograph observations by Philip C. Keenan and Paul Rudnick. Keenan was also to use the 24-inch reflector at Williams Bay to classify the brighter galaxies, in collaboration with Edwin Hubble at Mount Wilson, and a 6-inch ultraviolet instrument to measure their total brightness. There was also to be some spectroscopy and photoelectric photometry of galaxies by Struve, Elvey, and Keenan. Finally, Frank Ross would engage in photographic photometry in six of the areas at north declination 75 degrees selected by the Dutch astronomer Jacobus Kapteyn for international study of the structure of the Milky Way, and Elvey and Rudnick would use the photoelectric photometer to determine the brightnesses and colors of diffuse nebulae, following up on Elvey's observations of the Gegenschein. Frank Ross was continuing his optical researches, as well as obtaining Milky Way photographs at Yerkes, Mount Wilson, and Lowell observatories.

Clearly everybody was hard at work on contemporary classical astronomy, almost all in the stellar field, and with no sign of the kinds of unifying principles and insights which lay only a short distance in the future. A general review of contemporary astronomy throughout the world would have given much the same impression.

Struve then sketched an outline of his intentions for the McDonald telescope. A powerful coudé spectrograph, "similar to that used with the 100-inch reflector at Mount Wilson," would be used to obtain spectra of bright stars with very high dispersion. This instrument, together with a three-prism Cassegrain spectrograph, would be used to determine accurate radial velocities from high dispersion spectrograms. This would give important results for certain spectroscopic binaries and also for a "determination of the parallax of the Sun." There would also be a program of radial velocities of faint stars, in collaboration with other observatories. Because McDonald was farther south than any other observatory in the United States, a greater area of sky would be accessible to its instruments.

Because the focal ratio of the mirror would be 1:4, the instrument would be especially suitable for obtaining *spectrograms* of *diffuse nebulosities,* as well as of *spiral nebulae* (Struve's italics). "We shall use a quartz spectrograph at the principal focus for this purpose." Finally, "The McDonald telescope will also be used for other studies . . . : the positions, brightnesses and spectra of comets, the extension of our photo-electric work, the determination of photographic and photo-visual magnitudes of faint stars, etc. The large McDonald reflector will be suitable for various proper motion problems, such as the relative motions of bright stars with respect to faint stars or to distant spirals, and the internal motions of clusters and nebulae."

Thus wrote Struve in 1933, reproduced *in extenso,* since it is fascinating to see where he was right and where he failed as an accurate prophet, mainly because he did not foresee his own contributions to observational knowledge of many types of stars and to the theory of stellar evolution. For the time being, this was all in his head, and the matching of the dream to reality was to consume many years more. In all of this, there is no hint that any serious amount of work would be accomplished at McDonald before the large reflector came into service, in part because the director could not yet know how far in the future that would turn out to be. He was still assuring President Benedict in late 1935 that Warner and Swasey should be finished "before the winter of 1936."[54]

The same issue of *Popular Astronomy* carried another article pregnant with impending change in astronomy. Karl Jansky's "Electrical Phenomena That Apparently Are of Interstellar Origin" reported radio waves coming from a fixed region of the sky, without any indication that it was the direction of the galactic center. Jansky's discovery, which marked the beginning of radio astronomy, would eventually affect Otto Struve's post-McDonald career, when he would become director of the National Radio Astronomy Observatory.

McDonald Gets a Resident Astronomer

The first dwelling, House C, was supposed to be completed in the fall of 1934. Struve wanted to get a resident astronomer on the mountain as quickly as possible, partly to satisfy the Yerkes commitment to Texas, but also to take advantage of the southern latitude and the splendid transparency. Christian Elvey was the logical choice to go, but Franklin E. Roach was appointed instead, immediately upon receiving his PhD degree from the University of Chicago. Roach himself believes that his selection came because he was married, and Struve did not want to subject a bachelor to the isolation of a West Texas mountaintop.

The new resident arrived in Fort Davis in the midst of the town's eightieth anniversary celebration, on 6 October 1934, in a two-vehicle convoy together with his wife, Eloise, and their sons, John and Richard. They were accompanied by Theodor Immega, an engineer and recent immigrant from Germany, who was to be Roach's assistant and technician. The group was welcomed warmly by the celebrating citizenry and fed on a remarkable ranch chuckwagon stew known in polite company as "Son of a Gun." House C was not yet complete, so the Roaches rented a house in Fort Davis.

The workers were still in the early stages of construction of the large dome, so an interim research project was undertaken with a small telescope that the convoy had brought down from Williams Bay. Roach and Elvey had constructed a photoelectric device to measure the light of the night sky. Each observing night, Roach and Immega drove from Fort Davis to Mount Locke, where they would set up the telescope.

The telescope, which was apparently the personal property of G. E. Hale, was equipped with an altazimuth mounting, meaning that it rotated about the vertical axis and one horizontal axis. The photometer was set manually in altitude, the lens uncovered to activate the photoelectric cell, and then the rig was rotated in azimuth through an entire circle. The photoelectric current activated a galvanometer, which threw a light beam on photographic paper, which had to be moved in synchronism with the azimuth motion. Then the lens was covered, the altitude changed, and a new run could be made. The photographic records showed the variation of the light of the night sky, which not only is affected by the local weather and aurorae, but also measures the general glow of the zodiacal light and the Milky Way. Roach would develop the 8 × 10-inch sheets of photographic paper in the bathroom of the rented house the next afternoon.

"The drives to and from Fort Davis were rugged," Roach recalls. "The road was reasonably maintained, but Immega had never learned to drive, so even after a long night I had to fight sleep and maneuver the truck the 17 miles back to town. I told Immega that the least he could do was to keep talking to me, so I wouldn't go to sleep at the wheel. But he invariably fell asleep himself. Night after night, we would encounter two dips on the way home. I would hit number one at full speed and yell loudly to keep myself awake and also to wake up Immega. He would come to with a frightened start, but was invariably asleep again before we hit the second dip some hundred yards later."[55] (The road is now much less sporting since the recent construction of bridges over these branches of Limpia Canyon Creek.)

House C and a service building to house the diesel generator were eventually finished, at which time the resident astronomer really took up residence. Immega moved into the service building and could begin to sleep without fear of the dips. Living conditions remained pretty rugged, however. Roach suggested that the University of Texas rent a tourist cabin at Fort Davis one night per week, to allow the staff to take showers. This would be cheaper than hauling water at 1 cent per gallon.

In the late fall of 1934, a 12-inch refracting telescope was shipped from Yerkes and was installed on a temporary mounting on a spur of the mountain then known as Benedict Bench, the outcrop on which Houses A and B now stand. The instrument was later moved to a permanent pier on the western slope below the 82-inch building, where it remained until 1939. At first, it was used for making photoelectric observations of variable stars with the Yerkes photometer equipped with an old-style photovoltaic cell. Van Biesbroeck later used it for a variety of other work.

The continuing stream of newspaper stories attracted hordes of visitors, some of whom were incensed that they could not be admitted to the dome. It was still in the hands of the builders, but the tourists refused to believe that not even Roach had a key. To defuse this situation, he began holding open nights with the 12-inch refractor, with public lectures. This was embarrassingly successful and the lectures had to be transferred to the Fort Davis High School. Struve also permitted Roach to teach an astronomy course at Sul Ross State Teachers College, in Alpine, under condition that his name not be listed in the catalogue, lest it cause problems with the Chicago administration. This arrangement lasted for two years.

A Full Research Program Gets Underway

Christian Elvey was named resident astronomer-in-charge at Mount Locke in 1935, having acquired a wife just before his nomination. They arrived on the mountain on 10 November. The new resident was assigned to the McDonald staff, under the terms of the interuniversity agreement. The slot thus left vacant on the Yerkes staff was filled by the appointment as assistant professor of astronomy of a Dutch immigrant named Gerard P. Kuiper.

Struve was less enthusiastic about another aspect of the agreement, though. Elvey was obliged to act as local administrator, and he inquired of his director what title he should use in correspondence. The compact explicitly calls for Chicago to appoint a resident assistant director, but Struve demurred. "The President [of Chicago] is not in favor of any official title" he responded; Elvey must sign "for the Director."[56] There is no evidence that the president ever acted contrary to Struve's wishes on anything important during this period. There were going to be only one man's hands on the levers.

Roach remained to work with Elvey. The resident astronomical staff was completed by the arrival in September 1936 of Paul Rudnick and his graduate student wife Jessie (he got travel expenses, she didn't), and the June 1937 arrival of Carl K. Seyfert. To cope with the increased activity, a second engineer was hired to help Immega. Times were still hard, and there had been a large number of applicants, including one who had "studied religion, philosophy and psychology" and understood "the principles of astrological readings." He didn't get the job. It went to Arch T. Garner, the half-Cherokee who had worked as night watchman for Warner and Swasey until the steelworkers had arrived from Cleveland. When told he could have the job if his wife would run House E as a boarding house, he replied by postcard, "Will see you pronto." Eventually, Elvey promised him, there would also be a janitor/groundskeeper who could be trained to permit the engineers each to have one day off per week.

Under the terms of the agreement with the University of Texas, several persons resident at Williams Bay were also counted and paid as McDonald Observatory staff. For the 1936–1937 academic year, these included Struve himself (half time, director), Kuiper (half time, discovery of double stars from Mount Locke), Moffitt (full time, design of accessory instruments for the 82-inch and other telescopes), and machinist Charles Ridell (full time, instrument construction).

The year's salary budget for this galaxy of talent was $23,000.

What they had with which to mount a scientific program was the Yerkes 12-inch refractor, a solar photometer for measuring atmospheric extinction, a 6-inch ultraviolet telescope, and eventually the nebular spectrograph. A 3.7-inch f/2 Schmidt camera, intended for the Cassegrain spectrograph of the 82-inch reflector, was pressed into service as a detector.

The little camera was pivotal in the first scientific paper published in the *McDonald Contributions,* which also shows Struve's talent for improvisation. He and Elvey used the Schmidt to photograph various areas of the sky through the 12-inch refractor. Different filters were used to select colors of interest, primarily those associated with simple chemical elements such as hydrogen. Polarization was measured by holding a polaroid sheet in front of the telescope and rotating it to different orientations between exposures. In this way, they discovered nebulosities near Antares and in the constellations Scorpio and Ophiuchus that helped advance Struve's nebular reflection theory and permitted confirmation of the particle sizes in such clouds of dust and gas. To Struve, these findings represented "the most important advance in nebular research since Hubble's paper in 1922."[57]

Meanwhile, Elvey and Roach did photometric studies of spectroscopic binary stars, to find whether they showed variation in brightness. They also discovered an annual variation in the brightness of the zodiacal light. Elvey and the Rudnicks continued to examine the brightness and color of the night sky, now using the 12-inch and the Schmidt camera.

Carl Seyfert completed an ultraviolet study of eight large galaxies, discovering that the extended arms are significantly bluer than the nuclei, but not before receiving a severe reprimand from Struve that he should pay less attention to his horses and more to astronomy. Obviously, the message was heeded.

An observatory is stocked with human beings, after all, and the isolation that dark, clear skies require is always a potential stress factor. Some adapt, some crack, others just get grouchy. Theodor Immega got grouchy and was transferred to run the power station, so he wouldn't have to work with the others. By his own account, the problem was that an observatory is no place for an unmarried man. The transfer was a temporary expedient that in the long run didn't work, but that was largely due to pressures caused by an uncontrollable factor called World War II.

The Great Nebular Spectrograph

It was a tremendous nuisance not to have the McDonald reflector in service yet. There was a great amount of work to be done that could not be handled by existing instruments. In fact, it threatened to be more than a nuisance. The University of Chicago was getting restive about what it considered to be the low rate of return on its investment in the McDonald Observatory. Unless things picked up, the administration had threatened to reduce Chicago's part of the funding.[58]

Struve's first response was to suggest a whole new set of research programs that promised to give quick results. He listed among them: hydrogen line (H-alpha) photography of diffuse galactic nebulae, photoelectric light curves of "interesting" variables, infrared photography of red variables, and light scattering in dark nebulae. The latter would serve to test Struve's new theory of the reflection of stellar light within nebulae.

Elvey may have been on the mountain only "for the Director," but he would not roll over and play dead when the master spoke. Considering the state of the 82-inch mirror (estimates were still optimistic), it seemed pointless to him to begin new programs that would almost surely be abandoned when the new telescope came on line. If they were, "then you are going to ask us again in six months, 'what have you done?'"[59] The programs then underway were photoelectric observation of variable stars (one of the programs listed by Struve in *Popular Astronomy*), the color of the night sky, fluctuations of the zodiacal light, nebular problems of interest to Struve, and miscellaneous studies of extinction and scattering in dust storms. That should suffice.

Struve persevered. Good scientific reasons existed for pursuing the program that he had proposed. The driving forces, however, were internal politics at the University of Chicago. The three new *enfants terribles,* Kuiper, Bengt Strömgren, and Subrahmanyan Chandrasekhar, were producing such stunningly good work in such prodigious quantities that it was making the older astronomers look bad. Something had to be done quickly to restore some balance.[60]

A temporary compromise gave way a few months later to the pressures created by the lagging mirror progress. It was necessary to have an important instrument working on Mount Locke—soon. The 82-inch was not going to be it. Struve proposed that the Chicago astronomers put together a temporary device for the observation of the spectral characteristics of emission nebulae and other large and diffuse objects.

Elvey found a portion of the south face of the mountain that had a slope line pointing to the celestial pole, so that the local topography would supply the polar axis about which the light-gathering part of the instrument would rotate. Van Biesbroeck supplied the design.

Two piers were constructed on the slope, 75 feet apart. The lower one formed the emplacement for an equatorial head, which is a mounting whose polar angle can be set and fixed and which then can be driven around the polar axis to follow the sidereal motion of a fixed patch of sky. The main fixture on this mounting was a rectangular plane mirror of 36×10 inches, the light-gathering element of the spectrograph. The mirror was equipped with two parallel movable curtains to restrict its reflecting area. This "slit" of the spectrograph could be varied up to several inches in width. The upper pier supported a plane mirror of 24 inches diameter, facing down the slope. In operation, the lower mirror (called a coelestat) was pointed so that light from the region of interest was reflected to the upper mirror, which in turn reflected the light back down the polar axis to a small two-prism spectrograph. The dispersed spectrum was recorded on the f/2 Schmidt camera.

The movable mount carried two other devices necessary to the observations. One was a 4-inch telescope that looked at the region that was feeding light into the spectrograph. This required some clever gearing, since the principles of reflection required that the coelestat be pointed halfway between the direction of the target and the north pole, while the guide telescope had to point directly at the object.

There was another plane mirror, which fed a background sky

spectrum into the instrument, to be imaged alongside the nebular spectrum. This comparison enabled the observer to see which lines in the nebular spectrum were due to the nebula and which to the night sky. Research at McDonald and elsewhere had already shown that the sky often produces an emission spectrum somewhat akin to the aurora borealis, which could bias the nebular observations.

Since all of this optical paraphernalia was out in the open, the instrument had to have a whole series of light baffles and barriers designed to restrict the light falling on the spectrograph to that emerging from the nebula or from the comparison source. The observers even had to be careful not to reflect light from the sky into the spectrograph from their faces or clothing.

That was the McDonald Observatory 150-foot nebular spectrograph. Although technically it was a coelestat, in a sense it was rather like a telescope without the tube. Using the mountainside made one unnecessary. Discouraged by the progress in Cleveland, Struve opined that "we could probably use [it] for several years without feeling any need for the large reflector."[61] Over a period of three years, the spectrograph provided spectral data on many nebulae and other diffuse objects too faint or too far south to be observed with the same detector on the 40-inch refractor at Williams Bay, where it had been used for some time previously as a 69-foot instrument with an f/1 camera. In large measure, the two instruments were complementary and contributed to many of the same researches.

Jesse Greenstein, who was the first postdoctoral fellow specifically attached to the McDonald Observatory, had been using the Yerkes 69-foot nebular spectrograph for a study of the energy distribution in M31, the Andromeda nebula. When the detection part of the device was incorporated into the McDonald 150-foot nebular spectrograph, Greenstein went south to complete his observations on the new instrument. Analysis of the results of these data lowered the estimated temperature (or, strictly speaking, the intrinsic colors) of extragalactic nebulae in general and incurred the wrath of Edwin Hubble, because it upset Hubble's results on the numbers of nebulae of various apparent magnitudes.

Data from the McDonald nebular spectrograph were used by Struve, Elvey, and Walter Linke to determine the density of hydrogen in interstellar space. This led to Struve's discovery of high concentrations of hydrogen in some areas of the Milky Way, which in turn led to a confirmation of the recent discovery by Theodore Dunham at Mount Wilson Observatory of neutral calcium in the interstellar medium. The same observations were source material for several theoretical models of the interstellar medium, by Strömgren and others. Elvey and Linke used the instrument to continue the examination of the brightness and spectrum of the night sky, discovering that the sky contains occasional bright patches with enhanced sodium emission.

The giant remained in service into 1940, after which it was dismantled, having been rendered obsolete by the McDonald telescope.

The Case of the Disappearing Schmidt

Another pre-82-inch telescope continues to present a mystery. The Yerkes Observatory participated in the 1933 Chicago World's Fair, "A Century of Progress," in a spectacular and flamboyant way. The exposition was opened astronomically. A photoelectric detector attached to the 40-inch refractor at Williams Bay converted light from the star Arcturus into a pulse of electricity that was transmitted to Chicago to trip a switch to turn on the lights. Simple as this trick is, technically, it was a general marvel. Light from a star to light the fairgrounds; isn't *that* progress? Just in case of bad weather, the observatories of Harvard, Pittsburgh, and Illinois were ready as pinch-hitters.

The observatory secured a grant from the exposition committee, who had agreed that the astronomers' participation was so important to the fair's public relations that there should be some profit to them. Not monetary profit, but scientific profit. The fair would fund some new equipment. Among Struve's proposals was a 20-inch Schmidt telescope for the McDonald Observatory. This type of telescope, invented in 1929 by Estonian-German engineer Bernhard Schmidt, is a hybrid that uses a spherical mirror and a corrector lens to focus the image on a curved photographic plate located inside the tube. Its major advantages are that it has a very wide field and is very fast. The proposal was accepted, and work began early in 1938.

Bengt Strömgren, who wanted to use the instrument for photometry, provided the f/2.25 optical design, while van Biesbroeck designed the mounting. The construction of a dome to house it on Mount Locke is reported both in correspondence and in the Annual Report for 1937–1938.

Obviously, there were problems. Van Biesbroeck continued the design of the mounting in 1938–1939, and in 1939–1940, when it was reported that its fabrication had been completed by the Yerkes shop. Bills of lading exist for the rail delivery of the mechanical parts to Alpine, where they were to be offloaded for truck transport to the mountain. The Annual Report for 1940–1941 states that the dome was completed and the mounting installed, but that the optics would probably be delayed until the end of the war. Then, official silence. Little more is ever said about this instrument. Optical tests are reported in Chapter 11, of this book, but the telescope seems to have been left figuratively hanging in mid-air. No one now seems to remember anything about it. What happened to the optics? For that matter, what happened to the dome and hardware? Harlan Smith, present director of McDonald Observatory, recalls that, in the 1960's, there was a pile of rubble that may have included parts of the mounting. As for the rest, everything seems to have disappeared into the mists of the past, without a trace or a memory. Several former Yerkes/McDonald astronomers have suggested that the optics were eventually abandoned, but no records of either the act or the reasons have been found.

With or without a Schmidt telescope, astronomy had come to West Texas during the years 1934–1938. The primary mystery in everyone's mind concerned the world's second largest telescope. When was it finally going to be ready for use? What was going on in Cleveland?

The McDonald brothers in middle life: *left to right,* James Thomas, William Johnson, and Henry Dearborn.

Harry Yandell Benedict (1869–1937), Dean of Science,
later President, at the University of Texas at Austin.

Otto Struve (1897–1963), first director of McDonald Observatory. Yerkes Observatory photograph, University of Chicago, Williams Bay, Wis.

Worcester R. Warner. Warner and
Swasey Company photograph.

Ambrose Swasey lived long enough to see the completion of the mechanical
parts of the telescope. The two friends loved their sideline of telescope building at their machine-tool works. Warner and Swasey Company photograph.

Machining the central section of the telescope tube. Warner and Swasey Company photograph.

Finishing touches to the polar axis and driving gear. Warner and Swasey Company photograph.

Melting the Pyrex glass for the main mirror at the Corning Glass Company's works at Bradford, Pennsylvania. Warner and Swasey Company photograph.

Lundin and Burrell watch the main mirror disk on the surface grinding machine. Warner and Swasey Company photograph.

Franklin Roach during early days at Mount Locke.

George van Biesbroeck (*right*) with engineers on Mount Locke. The man in the middle appears to be Edwin P. Burrell. Photograph courtesy Harry Ransom Humanities Research Center, University of Texas at Austin.

Construction work on Mount Locke.

Support piers ready for the telescope.

The empty dome awaits the telescope. Warner and Swasey Company
photograph.

Christian Elvey awaits the telescope optics.

The upper part of the telescope tube en route.

The younger C. A. R. Lundin tests the mirror with the knife-edge apparatus.
Warner and Swasey Company photograph.

Robley C. Williams cleans the mirror before it receives its vacuum-deposited
coat of reflective aluminum. Warner and Swasey Company photograph.

J. S. Plaskett, C. A. R. Lundin, and engineer George A. Decker admire their reflections in the aluminized mirror. Warner and Swasey Company photograph.

The mirror arrives at Marfa. Edwin P. Burrell is standing by the crate.

The Yerkes staff, spring 1938. *Top row, left to right:* P. Nicholas, H. Sieghold, O. Struve (director), F. R. Sullivan, L. E. McCarthy, R. Wickham, O. B. Fensholt, C. Ridell; *middle row:* G. van Biesbroeck, Theodosia Belland, C. Hetzler, Frances Sherman, L. Henyey, Edith Kellman, Wm. Morgan, C. Rust; *bottom row:* Alice Johnson, S. Chandrasekhar, Mary Culver, B. Strömgren, Marguerite van Biesbroeck, G. P. Kuiper, Mrs. Chandrasekhar, J. L. Greenstein. Yerkes Observatory photograph.

Astronomers at the dedication of McDonald Observatory, May 5, 1939. Photo by Elwood M. Payne.

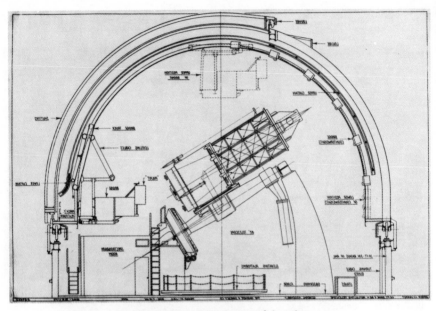

Sectional view of the optics and mechanics of the telescope.

Gerard P. Kuiper at Yerkes Observatory. Yerkes Observatory photograph.

Neptune and its two satellites. Top photo shows the planet and its largest satellite, Triton, taken on 24 February 1949 at the Cassegrain focus of the 82-inch telescope. Neptune is about 250 times brighter than Triton; notice the absence of other close satellites. This observation was only a few weeks before Kuiper's discovery of the second known satellite, Nereid. Bottom photo shows Nereid shortly after discovery, taken 25 May 1949 at the prime focus of the 82-inch telescope. It was necessary to expose so long because of Nereid's faintness that Triton, which is yet 250 times brighter, seems to be merged with the image of the planet.

The Austin faculty in 1963. *Left to right:* Harlan Smith, Gerard de Vaucouleurs, Robert Tull, Terence Deeming, and Frank Edmonds.

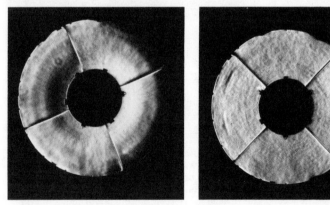

Knife edge tests of the Cassegrain system before refiguring (*left*) and after (*right*). Photograph by Jean Texereau.

Texereau refigures the secondary mirrors.

Marlyn Krebs (*left*) **and Jean Texereau.**

The 107-inch telescope.

The millimeter wave dish on Mount Locke.

Some of the helical antennae in one of the arrays of the University of Texas Radio Astronomy Observatory south of Marfa.

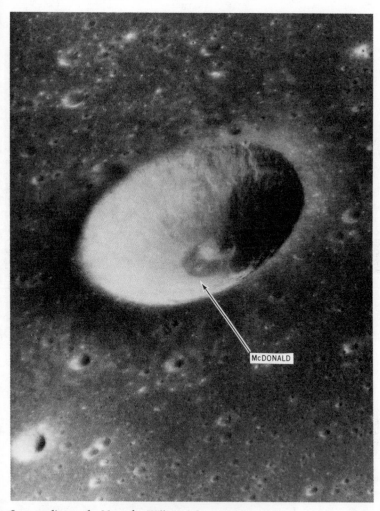

McDONALD

Immortality on the Moon for William Johnson McDonald.

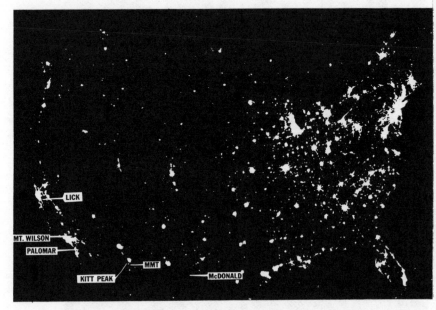

Night-time satellite photo, showing the locations of the six major continental U.S. observatories, all with telescopes of 100-inch size or larger.

CHAPTER 8

Mr. McDonald's Telescope

The regents of the University of Texas signed a contract for the biggest telescope they thought they could afford. The original concept was to have a principal mirror of about 81 inches diameter, which would provide an 80-inch clear aperture when mounted in the telescope. The aperture later changed by accident to 82 inches.

When the contract was signed in October 1933 with the Warner and Swasey Company for the design, fabrication, and erection of the McDonald telescope, all parties envisaged that it would be in operation by late 1935. The temporary expedients on Mount Locke in the mid-1930's, such as the loan from Yerkes of the 12-inch refractor and the ad hoc construction of the nebular spectrograph, permitted a respectable amount of astronomy to be done during this period, with staffing at about the levels that had been originally projected by the two universities. Nonetheless, these measures were necessitated by serious delays in the completion of the 82-inch reflector. It did not go into official service until 5 May 1939, although it had then already been used for several months.

The design of the McDonald reflector was innovative in many ways, and it is tempting to use that fact to explain the slowness with which it was built. Alas, life is neither so simple nor so benign. The innovative parts were ready on time and worked as expected. The delay was experienced in that part of the machine that, albeit extremely delicate, was perhaps most straightforward—the optics. Worse yet, most of the delay was the result of stubbornness and even incompetence, compounded by bad working conditions.

Optical Design

Contrary to what many people think, the prime function of a modern astronomical telescope is not to produce a magnified image of the sky, though this often occurs, but to gather light from an astronomical source. The telescope functions like a bucket put out in the rain of photons coming from a star. The bigger the bucket, the more rain is caught.

Telescopes range in design between two extremes. The type known as *refractor* performs all its functions by the use of lenses, which refract or bend the light going through them. The largest refractors are the 40-inch-diameter at Yerkes and the 36-inch-diameter at Lick Observatory in California, both of which were old when William Johnson McDonald died. They will probably remain the largest refracting telescopes, partly due to the difficulty and cost of making a large block of glass totally free from internal blemishes and then shaping it to required accuracy. In addition, there are severe mechanical mounting problems. Since the lens can be supported only on its edges, it is more prone than a mirror to distortion due to pinching and sagging under its own weight. Also, the lens must be mounted at the top of the tube, and the straight-through optical path requires very long telescope tubes that tend to bend. These features can impair the optical performance.

The type called *reflector* does it all with mirrors, as its name implies. The largest, or primary mirror, sits at the bottom end of the telescope tube, where it can be supported from underneath, as well as on the sides. The glass can have internal flaws, if they don't affect the strength and rigidity, because the light does not pass through the glass, but is reflected from its front face. The optical path is folded by reflection, so the tube is usually much shorter than that for a refractor of the same focal length. Since there is no refraction, the images are free from chromatic aberration, the dispersion in color that most lenses produce. Thus, not only can reflectors be made in much larger diameters than refractors, but for a given diameter they are much less expensive, and they avoid certain optical faults to which refractors are prone.

Consequently, one of the first decisions taken for the McDonald telescope was that it would be a reflector. With the money available, the University of Texas could buy the second-largest astronomical mirror that had ever been made, plus the telescope and building to go around it. Despite recurrent news stories to the contrary, the pride of Mount Locke would have no lenses as major components.

The word *refractor* nearly suffices to define an entire optical system. There is very little margin for design variants. The light goes through a lens at the top of the tube and comes to a focus at the bottom. Not so for reflectors. Once one begins inserting mirrors in the path, there is a seemingly endless variety of things that can be done. A complete specification of the optical configuration requires that the locations at which the light can be brought to a focus be defined. Four such focal positions were considered for the 82-inch telescope. They all require a parabolic primary mirror, and indeed can all coexist in the same instrument.

The simplest focal position is the *prime focus.* Here, the concave mirror, shaped like an oversize version of a shaving or makeup mirror, returns the rays of starlight back up the open-ended tube of the telescope in a converging beam that comes to a focus on the axis of the tube near the top end. The recording instrument must be installed inside the tube at that point. In some very large reflectors, there is room in the center of the tube for the observer to sit inside the telescope with his instrument.

With smaller telescopes, the prime focus must be accessed from a vantage point outside the tube. This is sometimes inconvenient or dangerous. To overcome this problem, a slight modification was devised by Isaac Newton for the first known astronomical reflector. Newton's problem was that his telescope was only two inches in diameter and only for visual observations. The prime focus was impossible to use, because the observer's head got in the way. The solution was to insert a small plane mirror just ahead of the prime focus, to divert the light beam out a hole in the side of the tube, making it more accessible. This is now called the *Newtonian focus.*

For any mirror diameter, the size of an extended image is directly related to the focal length, the distance the light travels from the primary mirror to the focal point. By inserting additional curved mirrors in the path, it is possible to extend the focal length beyond that for the prime focus. In the *Cassegrain* configuration, a secondary convex mirror is installed in the center of the telescope tube, just before the prime focus and facing the primary. This intercepts the converging beam of rays coming from the main mirror and sends it back down the tube with decreased angle of convergence, so that it comes to a focus just behind the back surface of the main mirror cell. The primary mirror and its mounting cell have central holes to let the light through. Nobody is sure of the identity of M. Cassegrain, the Frenchman who devised this scheme, but whether he was a mathematician or a sculptor his name is perpetuated in many an observatory.

Since the Cassegrain focus is at the bottom end of the telescope tube, it is of much easier access than the prime focus. Because it is much closer to the main support system of the telescope, much heavier equipment can be mounted here without risk of distorting the mechanism. Of course, the secondary mirror and its supporting fins block part of the aperture of the telescope. This fact and the idea that the main mirror has a central hole surprise many people, but in truth they do little beyond producing a slight reduction in the light-gathering power of the telescope. The secondary and the hole are in line, so that their effects are not additive. The fins cause only some minor peculiarities in out-of-focus images and spurious spikes on in-focus images of bright objects.

Even larger and heavier equipment can be used at the *coudé* focus. In this arrangement, the Cassegrain secondary is replaced by an even more convex mirror, which greatly reduces the convergence of the rays. These strike a small flat mirror mounted on the transverse axis about which the telescope tube rotates. The beam is sent out of the tube and into the hollow axis, after which one or more additional flat mirrors direct it down the axis to a fixed focal position in an optical laboratory, sometimes in a remote part of the building. Why *coudé*? It is a French word that means "bent in the form of an elbow." The light path goes through the double elbow joint formed by the telescope, declination, and polar axes.

The three configurations—prime focus (with Newtonian as an equivalent variant), Cassegrain, and coudé—have increasingly greater focal length and image scale. In consequence, they have successively slower optical speed. The choice of configuration depends on the work to be done. For example, the prime focus is used to observe very faint extended objects during the dark of the Moon. The Cassegrain focus is often used during the dark of the Moon for photometry of stars, photography, or spectroscopy of faint objects. The coudé focus is mainly used when the Moon is bright, most often for high dispersion spectroscopy of bright stars. A giant coudé spectrograph, far too cumbersome to be mobile, can show great detail in spectra.

Completion of the specifications for the McDonald telescope occupied a year of meetings and discussions in Chicago, Cleveland, and Austin, both before and after the contract was signed. Struve's original plans called for observing positions at the prime, Cassegrain, and coudé foci. At the recommendation of Dr. J. S. Plaskett, the respected emeritus director of the Dominion Astrophysical Observatory at Victoria, British Columbia, provision was added for a Newtonian focus. Struve did not think that it was really necessary, but it was included in the engineering drawings. It was eventually dropped for fiscal reasons. Because of recurrent folklore to the contrary, it is worth repeating: the first discussions between Struve and President Benedict in 1932 already envisaged a very long focal length coudé position for high dispersion spectroscopy.

Just at the point when finality of choice seemed to have been reached, a serious problem arose. Regent Lutcher Stark expressed doubts whether the proposed design and the choice of material for the mirror were the best available. He vigorously insisted on the new telescope configuration known as Ritchey-Chrétien, from the names of its American and French inventors. In fact, it developed that Stark was a friend of George W. Ritchey, which is likely to have influenced his opinions.

The science of optics is largely concerned with the fact that no optical system produces a perfect image of an extended scene in full color. One must choose the compromise that produces images that are the most suitable for the purpose at hand. In a normal telescope such as Struve wanted, the primary mirror has a paraboloidal shape, which produces the most perfect possible image of a star on the axis of the instrument. The images of objects off the optical axis worsen steadily the farther off they are. The Ritchey-Chrétien system is similar to the Cassegrain design, except that the primary mirror is not quite paraboloidal. By itself, it would produce only very imperfect images, but it is combined with a secondary mirror that matches the primary so as to give good focus over the entire field. The relative centering and orientation of the two mirrors is critical to their proper functioning, and this demands a very rigid and heavy support structure. It is also more expensive.

Struve countered the suggestion with an admission that the

Ritchey-Chrétien configuration was indeed an extremely important innovation, but that it was unsuitable for the research program planned for the McDonald Observatory. In addition, the stories that Stark had heard about deficiencies in the Victoria telescope were either exaggerated or untrue.

The obsession with the new design was only one of a series of roadblocks that Stark had thrown up in the path of Struve's intentions. He hadn't liked the idea of collaborating with Chicago to start with. He didn't think much of Warner and Swasey. The use of Pyrex glass was too radical for him, as was the new technique for coating mirrors. It appears that none of these objections came from blind prejudice, and some of them were based on reasonable grounds. He just carried them to extremes. Finally, on the telescope design issue, he went too far and was overridden by his colleagues. Board Chairman Beauford Jester reported to Chicago's President Hutchins that "Mr. Stark has made himself rather ridiculous in the eyes of the other members" and was no longer saying much about the affair.[62]

Lutcher Stark's reservations caused some concern at Warner and Swasey, who had their legal staff look into the force of his acquiescence rather than support. Did Stark have the power to dump the company at some later date? They were also concerned with the fact that Struve was privately desperate to keep the telescope within budget. In the end, Otto Struve got the design he wanted. Lutcher Stark was converted to a staunch supporter of the project, later becoming a very helpful chairman of the Board of Regents.

Mechanical Design and Construction

The principal component of an astronomical reflector is the large concave primary mirror. The light does not enter the material, but is reflected from a very thin metallic layer deposited on the front surface. The effectiveness of even an efficient collector of stellar radiation would be vitiated if the mirror did not reflect the light beams from each star to as nearly perfect a point image as possible. For this reason, the surface must be shaped with an accuracy of a small fraction of the wavelength of light. To maintain that shape under all climatic conditions requires that the material have a very low coefficient of thermal expansion, which means that it changes size very little with changing temperatures. For rather more than a century, nearly all mirrors have been made of some kind of glass or glass-like substance. Such mirrors are heavy.

The optician who grinds, polishes, and figures the mirror surface to its desired curve also wants that form to remain unchanged as the telescope is turned to different parts of the sky. This requires that the mirror be supported by a sophisticated system of pads, levers, and counterweights, which can be distributed all around the edge and back of the glass slab. Their purpose is not just to support the mirror in place, but also to keep it from bending.

As noted earlier, the telescope contains other optical components as well. The purpose of the telescope tube is to support them and keep everything properly aligned, no matter what direction the telescope is pointed. The apparatus that supports the tube must permit the instrument to be pointed to any part of the heavens and, once set, maintained accurately in place throughout the night, if need be. The mechanical arrangements that permit this are not difficult to understand.

The rotation of the Earth causes stars perpetually to move across the sky from east to west. The first task of the mounting is to counteract this motion. The easiest way to achieve this is to mount the tube on an axle pointed at the North Pole (or South Pole, in the southern hemisphere). This arrangement is called an *equatorial* telescope. When setting the telescope on an object, the tube can be turned rapidly around the *polar axis* until the proper position is reached. Then tube and axle are clamped together, and thereafter driven at a steady rate of 15 degrees per hour, just sufficient to match the rotation of the Earth on its axis. This keeps the telescope pointed at the same patch of sky.

Pointing in the north and south direction is achieved by mounting the tube on a *declination* fixed axis at right angles to the polar axis. The desired position is reached by turning the telescope around this axis. The tube and axis are then clamped together. The tube thus continues to point at a constant north-south direction, which is called declination in the sky.

Just as there are many variations in optical design, the mechanical requirement for two mutually perpendicular axes has been met in a variety of ways with different telescopes. Each has (or once had) advantages; each has disadvantages.

At the time that the McDonald telescope was being built, there were basically two viable options, given the current technology. They may be termed the *cross-axis mount* and the *fork mount*. The differences between them present both mechanical and astronomical choices. The fork mount, which was only then becoming feasible in its present form, features a single pier for the lower end. The declination axis is supported between the tines of a huge fork. The space for attaching equipment at the Cassegrain focus is small, unless the tines of the fork are made very long. They are then liable to deflection unless they are extremely heavy. In the cross-axis design, the polar axis is a single axle unit supported on two piers north and south, with the declination axis sticking out from a central box. This creates a balance problem.

Large telescopes are massive engineering structures weighing tens or even hundreds of tons. They must be moved through large angles with extreme precision. To avoid excessive loads on precise driving mechanisms, telescopes must be balanced. The tube must be balanced fore and aft. This is usually relatively easy, since the great weight of the primary mirror and mirror cell can be offset by a greater tube length at the other end and the lesser weight of the secondary mirror at its top. In symmetrical arrangements such as the fork mount, this is essentially the whole story.

For a cross-axis mount, with the tube on one side, the system must be balanced east and west too. The weight of the tube must be offset by another weight on the other side of the polar axis. In many cases, this weight is on an extension of the declination axis. If this extension is short, the weight must be very great, increasing the load on the lower pier bearing. This increases either construction cost or maintenance cost. If the extension is long, the weight is less, but the counterweight sweeps out a great deal of space as it is turned. It must clear the floor, and preferably the astronomer's head too.

Otto Struve wanted a telescope with observing positions at the prime, Cassegrain, and coudé foci. He also wanted a cross-axis mounting, providing the simplest access to the coudé focus.[63] This involved an element of risk with so large a telescope. The entire weight of the structure containing the optical system, some 13 tons, had to be supported on a short axis projecting from one side of the polar axis, imposing heavy radial and thrust loads on the bearings. Warner and Swasey's director of engineering, Edwin Burrell, achieved what he considered the "rather daring scheme" of cross-axis design by reducing the length of the declination axis and by using tapered roller bearings of a particularly large size.

By independent coincidence, the engineers of another telescope builder on the other side of the Atlantic chose the same solution to that same problem. When Grubb Parsons Ltd inquired of the bearing manufacturer for a guarantee of precision, the Timken Company asked how many thousand revolutions per minute the bearings would make. Of course bearings of a telescope might make part of a revolution in several minutes.

Burrell's solution to the counterweight problem was unusual. An arm attached to the polar axis was placed close to the tall north pier. It carried the counterweight always high above the floor, where it could not hit the elevator floor or slyly move into an unexpected position ready to brain an unwary observer in the dark. McDonald astronomers have been grateful ever since.

The Saga of the 82-inch Mirror [64]

The fabrication of the mechanical assembly presented no problems. The complete telescope, sans optics, was running perfectly in Cleveland by the end of March 1936. It was in place in its building on Mount Locke by the end of the year. Despite Struve's early optimism and the Warner and Swasey Company's promises, it would gather a lot of West Texas dust before any glass was mounted inside.

The mirror project had not even started auspiciously. In a daring move, Struve had decided to have his mirror made from the new Pyrex glass invented by the Corning Glass Works. It had a much lower coefficient of thermal expansion than ordinary glass, so the large diurnal temperature variation common in the southwestern states would cause less distortion. This move was opposed by Lutcher Stark, but Struve prevailed.

On the last day of 1933, Otto and Mary Struve and Christian Elvey were driven over icy roads to the Corning plant at Bradford, Pennsylvania, to witness the pouring of the 81-inch Pyrex disk. The hot slab went into an annealing oven to be cooled very slowly, so that it would not develop internal strains that might cause it to warp or crack during use.

When it emerged from the cooling oven over four months later, its surface was covered with fissures. One of them was an inch deep and five inches long. The Corning glassmakers assured Struve that this was nothing for concern. The "checks," as they called them, could be ground or sandblasted away. They had done this many times, with no ill aftereffects. The astronomer was aghast. "I undiplomatically described [them] as cracks," he later said.[65] At the time, what he said depended on his audience, but what he said to the Corning people was that the mirror blank must be remelted and recast. During the process, which took yet another four months, the pressure of the glass stretched the mold two inches. From then on, the McDonald instrument was known as the 82-inch telescope. Fortunately, the designers of the mechanical parts of the telescope had envisaged such a possibility, and the increased size presented no problem.

The task of transforming the huge piece of Pyrex into a precision mirror was entrusted to the young C. A. Robert Lundin. Warner and Swasey had hired him especially to obtain the Texas contract, their first venture into astronomical optics. Lundin had learned the optical trade as an apprentice to his father, also C. A. R. Lundin, in the shops of Alvan Clark. His father had been an acknowledged master, having figured the lenses for several of the giant refractors, including the record-holding 40-inch telescope at Yerkes Observatory. The young man had shown himself to be adept at refractive optics, but had never before made a large mirror.

The procedure for making a telescope mirror is long and arduous. The disk, or *blank,* is first ground to a circular form in an operation called *edging.* The front and back surfaces are then ground flat, to produce an exactly shaped disk. At this stage, it has a characteristic "ground glass" surface finish. At successive stages in the work, finer and finer grades of Carborundum and eventually jewelers' rouge are used to polish the surface. Extreme care must be taken to clean away every particle of one grade of abrasive before starting with the next finer one. Otherwise, remnant coarse particles can scratch the optical surface during the subsequent operation.

The reflective face is first ground to a spherical surface with a specially designed tool. This is first polished and then, with the utmost delicacy, *figured* by local polishing; figuring refers to the subtle changes of form needed to pass from spherical to the desired final shape. The polishing is done with a circular tool faced with pitch squares, using a polishing medium such as jewelers' rouge.

In the case of the McDonald 82-inch blank, Lundin calculated that a spherical surface required that the surface be undercut to a central depth of 1.314631 inches below the flat level, to obtain the desired focal ratio. The paraboloidal form required for the McDonald telescope was deeper yet by 0.001350 inches.[66] The accuracy desired was approximately one-millionth of an inch, all over the surface. This can be tested only by optical means.

Lundin had arrived in Cleveland in September 1933 to prepare for the great work. He did not have just one mirror to complete, but several. In addition to the 82-inch primary, there would be two 15-inch secondaries for the Cassegrain and coudé configurations, plus a 60-inch optical flat that would be necessary to test the primary.[67] These were ready for working sooner than the 82-inch blank, partly because of the remelting episode. Edging of these smaller mirrors was complete before the large one left the Corning Glass Works at the beginning of October 1934. Once the 82-inch blank arrived on 3 October, it too was soon trimmed to circularity. After that, Lundin was working on all the mirrors simultaneously.

Progress seemed to be quite satisfactory, as Lundin worked through the autumn and winter and into the spring of 1935. There was an accident in January, in which the arm of the 60-inch grinding machine gouged a chip 3 inches long and 1 inch deep from the back of the 82-inch disk; it was repaired by cementing a plug into the hole and smoothing it off. Despite this, the flat side was ground by 19 April 1935 and polished by 11 May.

Lundin started rough grinding what was to be the concave front surface of the 82-inch mirror on 15 May. By July, he was lowering the surface by about 0.007 inches in each eight-hour grinding session. At the end of that month, the rough grinding done, Lundin began the fine grinding with successively finer grades of Carborundum. This continued for two months, after which the mirror surface was ready for polishing.

The same machine was used for polishing as for the grinding, so it had to be cleaned with the most extraordinary care. It required four days to assure that no residue remained from the grinding operation. The tool was then coated with pitch, and polishing started on 15 October 1935. This is the point in fabricating a mirror when it ceases to be a purely mechanical process and begins to be an optical one. The surface is being shaped to conform to two characteristics: focal length and image quality. These must be tested frequently, and the polishing tailored to obtain the desired result. During much of the work, a test was conducted each morning.

The fundamental property of a paraboloidal mirror that makes it important to astronomy is that all parallel light rays from a distant source on the telescope axis are brought to a focus at a single location, the focal point. The focal length, which determines the length of the telescope, is the distance from the mirror to that point of convergence. The image quality is a measure of how nearly the mirror shape matches a perfect paraboloid, or how nearly the reflected rays converge at a single point. Good image quality can only be achieved if the real surface matches the ideal to within 10 percent of the wavelength of the light being studied. For the range of visible light with which optical telescopes work, this means that the optician should achieve an accuracy of about 0.000002 inches, or about 0.2 percent of the difference between the desired paraboloid and a sphere. Lundin wanted better than that.

For as long as he was working alone, Lundin evaluated the 82-inch mirror with a very sensitive technique called the *Foucault* or *knife-edge* test, which is capable of determining both focal distance and image quality. The basic principle of the Foucault test is that a perfect mirror will send all the light beams from a suitable source near the center of curvature to converge at a single point in the same general area. If a knife-edge is used to block the light precisely at that focal point, only an infinitesimal motion of the blade is required to pass from no blockage to total blockage. With his eye in a suitable position, the operator sees the whole mirror flooded with light. If he moves the knife edge in from the side exactly at the point of convergence of the cone of reflected rays, all of them will be cut off simultaneously and the whole mirror surface will go dark in an instant.

If the blade is before or behind the focal point, then there will be a zone in which the blade blocks only part of the bundle of light, because the bundle will have some finite thickness. As the blade is moved, an observer whose eye is quite near the blade sees a shadow pass progressively from one side of the illuminated mirror to the other. If the shadow comes from the same side as the blade, then the blade is too close to the mirror; if it comes from the other side, the blade is too far from the mirror. (See Figure 1.)

If the knife edge is mounted on a micrometer screw for motion toward or away from the mirror, the position of the point of convergence can be determined with extreme accuracy. A variation of the same system, but using light from a bright star, is frequently used by astronomers at the telescope to assure that their detector is at the focal point.

The first optical test at the center of curvature was made on 9 November 1935, when the radius measured 421.3 inches.

The mode of manufacture should produce surfaces that are rotationally symmetric, but an imperfect mirror may have circular zones of differing radius of curvature. This will also be evident in the Foucault test. Suppose, for example, that the central circle of the mirror has one curvature and the outside zone another. These will produce two different zones of ray convergence, one from the outer part of the mirror, one from the inner part. If a knife edge is brought in at a point intermediate between the two distances, one ring will go dark from the right and the other from the left. By adjusting the position of entry of the knife edge, the separate focal points can be determined, in the one case by concentrating attention on the mirror's central zone, in the other by concentrating on the periphery. (See Figure 2.)

This was the test principle that Lundin was using to guide his figuring of the 82-inch mirror. It is a very good principle, still used by the finest opticians. One must, however, use it correctly.

The work went very slowly from the time figuring began. Struve had assured President Benedict when Christian Elvey was dispatched to Mount Locke that "I see no reason why the Warner and Swasey Company should not be able to complete the 82-inch telescope before the winter of 1936." [54] The figuring was still underway in September 1936, and Lundin refused to give an estimated completion date. Struve privately estimated one to six months. [68] The Yerkes/McDonald director began to suggest another sort of image quality test, the Hartmann, but Lundin wanted none of it. Struve was clearly uneasy about progress.

The working conditions in the Warner and Swasey plant were far from ideal. When under test, the mirror was supported in a vertical position on an 8-inch steel band. This showed signs of distorting the surface, producing elongated images. The tempera-

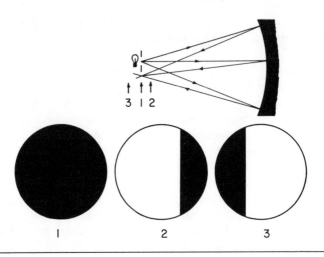

Figure 1.

ture control of the optical shop was inadequate, with steam pipes and heaters close by the mirror. Lundin was also troubled by vibration of the building, caused by nearby machines and trucks. He often worked on weekends, when the industrial activity was shut down.

A few months later, Warner and Swasey management was uneasy, too. At one point, even the director of engineering (not an optician, but a mechanical genius by all accounts) had tried to help in the figuring work, and had only made the figure worse. The solution was to hire as consultant Dr. John Stanley Plaskett, emeritus director of the Dominion Astrophysical Observatory. He had used a variety of techniques to test the 72-inch mirror in the Warner and Swasey mounting there, so the company knew him well. Plaskett had advised Struve on many aspects of the McDonald telescope design, and had called his own expertise to the attention of the Warner and Swasey Company as early as 1933.[69] He took over responsibility for the mirror figuring on his first official visit to Cleveland, in April 1937. "Lundin estimates an August finish,"[70] Struve reported to Elvey.

There was no August finish. In October, Struve visited Cleveland again, and he came home distinctly worried. "I see no material progress since last February. They are talking about the possibil-

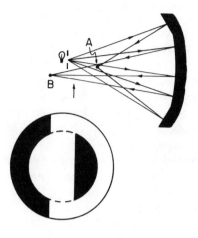

Figure 2.

ity of finishing it in November, but I am not at all certain that Lundin will be able to carry out this schedule."[71] Then some good news. In November "Plaskett informed me . . . that in his opinion the mirror will be ready for tests about January 1."[72]

There were no final tests in January, and Struve returned from Cleveland in March furious. For the first time, he and van Biesbroeck and Kuiper had been shown every detail of the procedure, and it was wrong! Lundin had misinterpreted the meaning of the Foucault test and was taking material off the mirror in the wrong places. Basically, the problem was that a displacement of the focal point indicates a bad angle on the mirror face, and Lundin treated it as though it were a bump of excess material to be removed. Consequently, he was removing material in the wrong place and merely transforming one deformation into a different one. It was now clear to the visiting astronomers how the optician could have the figure accurate to one wavelength and the next day, after an "inspiration," it was degraded to ten as the result of a misunderstanding of the Foucault test.

It is possible, but far from certain, that the error arose from the fact that Lundin had previously worked only with lenses, rather than mirrors. The 82-inch mirror was probably the first nonspherical surface that he had ever figured. In any event, Struve regained his good humor. As he said to Elvey, "I am really looking forward to a speedy completion of the mirror." And then he reflected, "But what are we to think of Mr. Lundin, and especially of Mr. Plaskett?!"[73]

There was no speedy completion. The next month showed no progress. No change seems to have been made in the regimen. Exasperated, Struve sought and received the backing of the University of Texas to demand a "change of management." He demanded the dismissal of Lundin and the installation of Yerkes optician E. Lloyd McCarthy as scientific supervisor of the optical work. Warner and Swasey vice-president Charles J. Stilwell temporized on the first request, but acceded to the second. A decision on Lundin was deferred for a month. McCarthy took up a permanent, although informal, presence in the shop. Plaskett was apparently in poor health and went home to Victoria, not to return to Cleveland until the mirror was ready for its final tests.[74]

Struve clearly was convinced that the fabrication of the 82-inch mirror had pushed both Lundin and Plaskett to their Peter Plateaus,[75] that point where the job was beyond their capacities. No longer willing to trust either one of them to produce a satisfactory result, Struve was relieved to have one of his own Yerkes people keep an eye on what they did. Indeed, from then on the progress was rapid, and there were no more discussions of personnel changes.

As if the figuring problems were not enough, Lundin was also having to cope with another new technology. The high reflective efficiency required of astronomical mirrors is achieved by coating the front surface with a thin layer of metal. Up to the mid-1930's, the customary method was chemical deposition of silver, but a technique had been developed for vacuum deposition of aluminum. This resulted in a more uniform layer of metal. In addition, aluminum is more durable than silver, and a superior reflector in the ultraviolet region of the spectrum. Otto Struve had proposed to use this method very early on, and the Texas regents had approved it—over Lutcher Stark's objections. In fairness to Stark, it was a radical idea at the time, but he eventually became convinced of its correctness and supported the proposal when it counted.

Lundin began experimenting with vacuum deposition in mid-1936, but he had his problems. That was not what he had been hired to do, however, so Warner and Swasey called in one of the developers of the process, Dr. Robley C. Williams of the University of Michigan, to oversee the final coatings. Williams arrived to work on the vacuum chamber early in 1938.

Everything was now going well. Some smaller mirrors were aluminized in August. Final figuring was done in September, and Otto Struve signed acceptance of the 82-inch mirror on 15 October 1938. The event was celebrated with dinner at the University Club. It had been a long, hard road for all. They had not yet reached the end, but the goal was finally in sight.

First Light

As soon as the big glass had been aluminized, Plaskett undertook the final Hartmann tests to evaluate the precision with which the ideal paraboloid had been realized. This test is accomplished by covering the mirror with a stiff diaphragm which is pierced with a radial pattern of holes. All of the holes at a given radius from the center of the mirror will reflect the light rays that fall on them to the same focal point. If the light from two symmetric holes is photographed from inside the focal point, the picture will show two dots, separated by a distance governed by how far the camera was from the focal point of the mirror. The same operation from outside the focal point will also produce two dots, with the images reversed. The distances between the images on the two photographs, along with a measurement of the two positions of the camera, permit an easy calculation of the location of the focal point for that zone of the mirror. Repeating the operation for each zone of the mirror gives the figure of the mirror.

When the results of Plaskett's tests were in, a press release declared that the mirror was perfect to 5 percent of a light wave, or about one-millionth of an inch. It was, so Plaskett was quoted, the most nearly perfect mirror in the history of the world. More reliably, it was noted that the giant weighed 5,600 pounds, was 82.3 inches in diameter, 11.63 inches thick at the edge, and had a central hole 13.35 inches in diameter.

Later experience indicates that Plaskett was optimistic. In 1962, Struve said that "the final tests . . . indicated an excellent figure, but this work was done in the shop and not at the telescope. The fact is that the secondary mirrors were tested quite superficially. The 82-inch mirror has an excellent shape, but its original supporting mechanism frequently allowed it to become astigmatic."[35] Two years after that statement, optician Jean Texereau came from France to Mount Locke to refigure the secondaries, and his opinion is quite different. Commenting recently on his experience, he remarked, "With a Hartmann test, one can show what he wishes, if it is not controlled visually [with a Foucault test]."[76] He found errors of nearly a wavelength in the primary, there from the beginning. At certain temperatures, the figure is briefly quite good, and the Hartmann tests may have been performed under this condition.

The mirror arrived at the Alpine railroad station before daybreak on 22 February 1939. Struve hoped it would arrive with no fanfare to arouse public attention. Assuming that a certain lunatic fringe thinks that shooting big mirrors is even more fun than riddling highway signs, the Warner and Swasey Company had enclosed the 7-foot disk in a large case, as had already become standard precaution. The bulky but otherwise unobtrusive coffer was offloaded from its boxcar, transported to the mountain, and hauled by manpowered hoist up the outside of the building and into the dome. The foreman gave the hoist ring, machined from a single piece of plate, to Keesey Miller. Miller wanted it as a frame for a portrait of his father, who had done so much to secure the observatory for Fort Davis.

March came in like an astronomical lamb. The disk of Pyrex was mated with its telescope on the first day, in the presence of a large audience that included Struve, van Biesbroeck, Elvey, Kuiper, Stilwell, Plaskett, and others. That night, the knife edge was used at prime focus to look at Sirius. For Plaskett, "It was a thrilling experience to see the shadow pattern from a star for the first time." First light had come down the tube and been reflected to an observer. After this test, the knife edge was replaced by an eyepiece, and the gathered assemblage amused themselves by looking at the Moon and Orion nebula. Plaskett called the images "beautiful."[77]

Van Biesbroeck and Kuiper took the first direct photographs at prime focus the following night, with the Pleiades and the Orion nebula as targets. They were not in focus. The Yerkes-built knife edge was found to be slightly out of the plane of the camera plateholder. This was quickly adjusted, and satisfactory photos were obtained on subsequent tests.

On succeeding nights, the Cassegrain and coudé secondaries were installed and tested. On March 5, Struve used the Cassegrain spectrograph to observe the peculiar star 17 Leporis; the ultraviolet spectrum was later discussed in *McDonald Contribution No. 14*, the first scientific results of observational material obtained with the new telescope. It may be significant that a telescope's optical figure is not overwhelmingly important to spectroscopy.

Plaskett reported to Warner and Swasey's President Bliss that Struve had pronounced the images "perfect." "I firmly believe," Plaskett continued, "that [the optical equipment] is unequalled in any existing telescope."[77] Charles Stilwell, Bliss's vice-president and soon his successor, was thrilled by the experience of those nights on the mountain: "There are not many experiences of that kind for the average businessman, but when they occur they make the humdrum of our commercial lives seem more worthwhile."[78]

This same man would say of astronomy, some sixteen months later, "I know of no other field of scientific endeavor in which industry fails so completely to secure recognition at all commensurate with the contribution which it makes."[79] The Warner and Swasey Company had, in fact, exerted itself well beyond the line of duty and had lost $85,000 on the deal, not counting used shop equipment donated to the observatory, nor the considerable expenses associated with the dedication. The mechanical design and the building had been executed superbly. As far as the public knew, the optics had been, too. The public isn't always told everything. Even after he had pronounced the telescope performance to be perfect, Struve was not satisfied—and with cause. Warner and Swasey's expansion into the field of large optics had not been fully successful.

One week after Struve's glowing comment, he informed Elvey that the "problem of deformed images" might delay formal acceptance of the telescope by the University of Texas until after the dedication ceremony.[80] That surely would have thrown a wet blanket on the party, and everyone was happy that it didn't come to pass that way. The trustees of the McDonald Observatory Fund accepted the 82-inch telescope for the University on 25 March 1939.

At long last, William Johnson McDonald's telescope was ready to look beyond the gates of heaven.

CHAPTER 9

The Dedication

Thirteen years after the bequest and several years later than Struve had hoped, the telescope was a reality. Chicago astronomers, the regents of the University of Texas, the Warner and Swasey Company, and the inhabitants of Jeff Davis County were united in a determination to celebrate in true Texas fashion.

Despite their fiscal loss on the project, the builders offered to underwrite most of the expenses for the dedication ceremony. There was to be a formal transfer ceremony, a real ranch barbecue, and a three-day scientific symposium. President Morelock arranged for the entire affair to be preceded by a three-day regional meeting of the American Association for the Advancement of Science, held on his Sul Ross State College campus, 40 miles distant from the observatory. The last day, 4 May 1939, was to be on astronomy, with an evening talk, "Physics Views the Future," by one of the deans of American physics, Arthur H. Compton.

Invitation lists were compiled in several stages. Of course, there were the professional astronomers. There weren't many in those days, perhaps only 10 percent of the current astronomical population. The invitees included nearly all of the astronomers of any standing in the world. Not all of them could come, of course. West Texas was distant, and many could not afford the time. A surprising proportion did attend, however, and they probably represented the most remarkable galaxy of astronomical talent gathered together in any one place in the years between the two world wars.

Other lists of prospective invitees included the presidents of thirty-one major universities, many state officials, the entire Texas congressional delegation along with Vice President John Nance Garner (a native of Uvalde), over a dozen prominent Texas amateur astronomers, and the news media. The Chicago administration provided a list of past or prospective benefactors, including officers of several foundations, beer baron Adolphus Busch III, drug maker Eli Lilly, and newspaper power Colonel Robert McCormick, apparently on the principle that there is nothing like a good example. One of the lists shows a curious Chicago mentality: three pages of names from all over the state of Texas were labeled "Local"!

Alpine high-schooler Calvin Richmond wrote to Elvey asking for an invitation, noting that he and his father were building a 6-inch telescope. The astronomer responded with a formal invitation to the dedication ceremony, but not to the symposium. He explained that the scientific meetings would be at a professional level and therefore would probably not be very interesting to the teenager. The public open night would be far more appropriate, said Elvey, and he sent an admission to that also. Young Richmond returned the public night ticket with thanks. "In view of the fact that we are local folks and can see the telescope at another time, I am returning [the ticket], that some out-of-town person might be favored."[81]

Getting There Is Half the Fun

The Warner and Swasey Company took no chances that the dedication would fall flat from lack of astronomers. Special trains, or at least special groups of cars, were chartered by the telescope manufacturers to transport the scientists to Alpine. They came via the Illinois Central, the Frisco Lines, the MKT, the Rock Island Line, and the Southern Pacific.

The atmosphere on board may be judged by the printed menus that were provided for all meals, particularly those of the Southern Pacific. Each one was graced by a cover photograph of some landmark served by the railroad: the Alamo, Carlsbad Caverns, the Papago Mission near Tucson. The inside cover shows Christian Elvey contemplating his giant new reflector, with the menu on the facing page. They dined well, even if the names of the dishes are amusingly obscure. Is there a cookbook somewhere that describes the preparation of Broiled Lambchop Gaposchkin? How about Kippered Herring with Scrambled Eggs Bobrovnikoff, or Breaded Veal Cutlets Curtis Style, or Grilled Breakfast Steak à la Mrs. Chandrasekhar (!), or Boiled Halibut Hollandaise à la Stebbins . . . ?

One of the younger astronomers on the invitation list was Peter van de Kamp, who was also fascinated with the relatively new hobby of motion picture photography. The opening scenes of his movie of the McDonald Observatory dedication capture the spirit of the railroad age with remarkable power and art.

Once arrived, the participants were housed in virtually every available room in the area. The presidents of the University of Chicago, the University of Texas, and the American Astronomical Society were lodged with the Struves in House A. Eight people (including three couples) were housed in the telescope building itself. Three chose to camp out on the site. The Prude Ranch, a dude ranch (and working ranch as well) 4 miles on the McDonald side of Fort Davis, was full. So was the West Texas pride of the Roosevelt administration's Work Projects Administration, the Indian Lodge in Davis Mountains State Park. W. S. Miller's Limpia Hotel overflowed, as did all the other commercial inns for miles around. In the face of unprecedented needs, the citizenry responded magnificently. Ranchers and townsfolk offered spare bedrooms to total strangers, often foreigners, come to celebrate the new observatory. West Texas is like that.

The evening of 2 May, the first day of the AAAS meeting in Alpine, there was a formal public night and demonstration of the telescope. Espy Miller organized some of his friends, dressed up in "the cowboy outfit" for local color, to handle the crowds. Over 250 visitors came, were assembled into groups of 50 or less, and were guided through the dome, over to House A for refreshments, and then back down off the hill. It was necessary to keep things moving to assure that latecomers could find parking space.

D Day

Finally, it was Friday, 5 May 1939; the long-awaited day had arrived. The participants were transported by bus to the observatory, where the ceremonies were to be held on the observing floor of the telescope dome. There, they found Elvey at the mechanical controls of the telescope and observing stations, Seyfert running the projector, and Kuiper serving as scientific aide to the press.

Henry Norris Russell presided over the morning session, which Harlow Shapley opened with "Recent Advances in Astronomy." He was followed by J. Gallo of the National Observatory of Mexico, who described astronomical work in his country. The morning wound up with a description of the telescope by J. S. Plaskett. Telescope domes are not renowned for their acoustics. Plaskett, speaking on the technicalities of mirror testing from the top of the coudé spectrograph housing, was alternately inaudible and incomprehensible.

For the official participants, lunch break was a full-blown chuck-wagon barbecue dinner and rodeo at the Prude Ranch, compliments of the Warner and Swasey Company. Earlier, company treasurer John C. Kline had remarked on the cost, but had added that "Charles Prude understood that [Charles Stilwell] wanted a real show, and . . . we will get one."[82] The astronomers were treated to several regular rodeo events, such as cattle roping, bronco riding, and calf bulldogging in an arena improvised from a row of parked automobiles. There was also a parade of "old-timers" and a Mexican band.

The formal dedication ceremonies resumed at 3:00 P.M., along with the arrival of bad weather. Most observational astronomers are inured to such perversities of natural phenomena. Nonetheless, one participant wrote home with the comment that "it is said to rain here only five days out of the year, and we have already had 60% of this year's quota in the past week."[83]

The spirit of the occasion was dampened less by the climate than by the absences, however. Several of the most important contributors to the success of the project were unable to be there. Ambrose Swasey had been dead for several years. Texas President Harry Yandell Benedict and Warner and Swasey's Edwin P. Burrell had both died in 1937. The company president, Philip E. Bliss, had died unexpectedly only three weeks before the dedication, occasioning the additional absences of his wife and Mrs. Burrell, who stayed behind to comfort the new widow. The four were eulogized in the course of the afternoon by Texas President *ad interim* J. W. Calhoun and by Charles J. Stilwell.

The regents of the University of Texas were present in force. Regent Edward Randall presided at the dedication session. Also present were Regents Kenneth Hazen Aynesworth, Marguerite S. Fairchild, H. J. Lutcher Stark (now a firm supporter of the observatory), and the president of the board, Jubal R. Parten. Chicago President Robert M. Hutchins was ill, and his message was read by Dean of Sciences H. G. Gale.

The observatory was formally tendered by Stilwell, who spoke at some length of the history of the Warner and Swasey Company and its founders. Stilwell and his colleagues had lived with this observatory for six years and had appreciated the friendship of such persons as Lutcher Stark, President Calhoun, and the West Texas ranchers. The McDonald Observatory would be in good hands with Otto Struve and the two collaborating universities. "In the Book of Hebrews, there is a passage 'Every house has a builder, and the builder of the universe is God.' On behalf of the Warner and Swasey Company, I hereby give to you the house we have builded. Our hope is that it may be used to show us more of God's Universe." [84] In his own way, Charles Stilwell was also pointing W. J. McDonald's telescope through the gates of heaven.

Acceptance

In his acceptance speech, Struve addressed the goals of the Mc-Donald Observatory in conciliatory tones. "Aperture is important. For some problems it is all-important. These problems are not within the scope of our observatory. We shall not attempt to extend the boundaries of the universe of galaxies: this task is taken care of by Dr. Hubble at Mount Wilson. Nor shall we attempt to photograph stars within our own galaxy which are fainter than any hitherto recorded. Dr. Baade is engaged in this type of work with the 100-inch." [84]

"What we propose to do," he continued, "is to study intensively the relatively bright stars of our galaxy—as individuals and not as statistical material. We want to know why it is that all matter in the world is segregated essentially in two forms: stars and nebulae. Why are there no stars which exceed in mass a few hundred times the mass of the Sun? Why is it that nearly all stars and nebulae consist of the same chemical elements in roughly the same relative proportions as we find them in the Sun? Where and how do the stars generate their stupendous energies of light and heat, and what is the ultimate fate of their radiation?"

Astronomers now think they know the answers to most of these questions. Struve showed extraordinary prescience by asking them in 1939.

Jubal Parten, in a graceful speech of acceptance for the Board of Regents, gave a biographical sketch of the benefactor. J. S. Plaskett spoke once again about the telescope optics. Arthur Compton, in an essay on "The First of the Sciences," waxed with that pomposity that then still occasionally passed for eloquence, with quotations from Macaulay, Lucretius, Francis Bacon, Alfred Noyes, and University of Chicago poet E. H. Lewis.

Finally, U.T. President-elect Homer P. Rainey spoke with approval of the thirty-year agreement with Chicago, pledging the complete cooperation of the University of Texas. "We are here to dedicate the Observatory to the most ancient and purest of all the sciences. [May it] stand as a symbol of the freedom of man's mind to explore the boundless areas of truth without any restrictions whatever. To these ideals, I dedicate the McDonald Observatory in the name of the University of Texas, and I now declare it open for research."[84]

This last was a mild bit of hyperbole, however excusable. The McDonald Observatory had been open for research for most of five years, and even the McDonald telescope had been in use for two months.

The Milling Scene

There is no way to be sure how many participants there were at the dedication. How many doesn't really matter. What matters is who they were and how they viewed the event. At least two people, Peter van de Kamp and Sarah Kuiper, made home movies. Several, including C. A. R. Lundin, Frank Ross, and J. M. Kuehne, gathered autographs in their personal copies of the little book *The McDonald Telescope,* printed and distributed by the manufacturer. Harlow Shapley's copy is in the library of the American Philosophical Society. The composite "group picture" of the affair assembled from these materials is a microcosm of recent astronomical history. Aside from those already mentioned, it included Walter S. Adams of Mount Wilson; Russell W. Porter, renowned polar explorer and design engineer-artist on the Mount Palomar 200-inch telescope; E. F. Carpenter, who later discovered the first known flare star; Edwin Hubble, the one-time lawyer cum astronomer who discovered the expansion of the Universe; and Jesse Greenstein, to whom Hubble refused to speak, because of his recent work on nebular spectra.[85]

There were also Carl K. Seyfert, who later gave his name to a new class of energetic galaxies; Albert E. Whitford, pioneer of photoelectric photometry; Walter Baade, originator of the concept of stellar populations; Henry Norris Russell, then the dean of stellar astronomy; Harlow Shapley and Heber D. Curtis, protagonists in the Great Debate on the nature of galaxies, held in April 1920 (neither was entirely right); outstanding optical designer and astronomer Frank E. Ross; and Subrahmanyan Chandrasekhar (whose name translates as "He Who Carries the Moon"),[86] who revolutionized the mathematical theories of stellar structure and evolution and of celestial dynamics, work that eventually brought him the 1983 Nobel Prize in Physics.

There were a host of other great astronomers: Albrecht Unsöld, E. A. Milne, Bertil Lindblad, Joel Stebbins, N. T. Bobrovnikoff, F. S. Hogg and his astronomer wife, Helen Sawyer Hogg, Cecilia Payne-Gaposchkin and her astronomer husband, Sergei Gaposchkin, Jason Nassau, Bart Bok, and many more. Reproductions of the autographs are included elsewhere in this book.

In Peter van de Kamp's film, one can occasionally spot a very young and furtive-looking Martin Schwarzschild. He wasn't on the invitation lists. When taxed with this, Schwarzschild admits that he crashed the party, but innocently. Imagining the dedication to be open to the public, he came at his own expense and without reservations. On arriving, he discovered his error and simply decided to bluff it out and see if he would be thrown out. He wasn't. "I can't imagine how I could have done a thing like that," he says now.[87]

The pictures and reminiscences include a number of people who were not identified at the time, some of whom remain anonymous. They were sometimes listed as "unknown journalists." One who suffered that fate later became resident astronomer of the observatory.

Regrets and Best Wishes

Many of those who could not attend sent letters or telegrams of regrets and best wishes. Foremost among them was Sir Arthur Eddington, perhaps the most eminent astronomer of the time. He had originally been scheduled to keynote the meetings on Mount Locke, but was unable to come.

The spring of 1939 was a tense time on the political scene, and several congressmen sent their regrets. In a characteristic message, only one took the trouble to explain that he was very sensitive to the value of astronomy, despite the fact that the international situation required that he remain in Washington; first-term Con-

gressman Lyndon B. Johnson was already adept at what has come
to be called "stroking" his constituency.

A unique communication came from the astronomers at Florence, Italy. It is a highly decorated colored scroll inscribed in
Latin. Translated, its message reads:

> From the hills sacred to the memory of Galileo, who was the
> first man to scrutinize the stars with a small telescope, the
> astronomers of Arcetri join in greeting at its inception a New
> World observatory that will reveal the paths of the universe.
>
> Dated at Florence five days after the kalends of May 1939,
> the 17th year of the restored Fascist era

Strictly, this means the sixth of May, but maybe the Florentines
were a little hazy about the exact date of the dedication. The
document was signed by several distinguished Italian astronomers, among them Giorgio Abetti, Atilio Colacevich, Guglielmo
Righini, and Mario Fracastoro. Possibly because of its alternative
system of dating, this beautiful document was found in bad condition by one of the present authors, tucked away in a file cabinet.
It has now been restored to the respect that it deserves, as a result of the research for this history.

The Problems of Mass Communication

It was clear virtually from the moment that Franklin Roach arrived on Mount Locke in 1934 that there would be a high level of
journalistic interest in the doings there. This has not changed in
half a century. Few moments have been so charged with repor-
torial excess, though, as those days in May 1939 when the eyes of
the astronomical world—and many of its most important bodies—
were concentrated on that half-nude mountain in West Texas.

In 1939, motion picture theaters were in their heyday, and the
short newsreels served the same purpose for many people that
the television news does today. The Public Relations Office of the
University of Texas transmitted the following telegram to Elvey
on 2 May 1939, just three days before the main event: "James
Lederer Paramount News arrives Wednesday . . . making pictures
[to] be shown to 80 million people." The PR director, William
McGill, made it clear that this was important and that Elvey should
cooperate in every way possible or reasonable. There are no de-

tails of the aftermath, but the few hints that exist are more than suggestive.

Telegram typescript, no date, to Paramount News, New York: "Recall James Lederer Austin claimed to represent Paramount News at McDonald Observatory dedication Friday May 5th. Your cooperation greatly appreciated. [signed] Otto Struve and Warner Seely."

Telegram typescript, no date, to Paramount News, New York: "Newsreel photographer covering McDonald Observatory dedication included girl sequence objectionable to observatory directors and sponsoring educational institutions. Feel sure you will agree this sequence has no place in this science coverage. Will appreciate its elimination. [signed] Otto Struve, director, and Warner Seely."[88]

Defining the Starting Point

With Dedication Day over, the Warner and Swasey folk (and various others) could relax. They headed for the Chisos Mountains, in what is now the Big Bend National Park.

The astronomers still had work to do. The symposium on Galactic and Extragalactic Structure continued through the entire weekend, evenings included, and half of Monday, 8 May. The scientific program was a remarkable summary of the state of astronomy, with the presentations made by those who had already or soon would become among the world's authorities on their topics. The only way to avoid a massive overuse of superlatives is simply to list the program.

> J. H. Oort, "Present Problems"
> R. J. Trumpler, "Star Clusters"
> Otto Struve, "Interstellar Matter"
> Bart Bok, "Star Counts"
> C. T. Elvey, "Galactic Light"
> Gerard P. Kuiper, "Under-Luminous Stars"
> E. A. Milne, "Cosmological Theories"
> Walter Baade, "Photometric Problems"
> Harlow Shapley, "The Space Distribution of Extragalactic Nebulae"
> Cecilia Payne-Gaposchkin, "Stellar Spectra and Colors"

W. W. Morgan, "Stellar Colors and Luminosities"

Bertil Lindblad, "Theoretical Interpretation of Spiral Structure"

S. Chandrasekhar, "Star Streaming and Dynamics of Stellar Systems"

Edwin Hubble, "Structural Features of Extragalactic Nebulae"

Joel Stebbins, "Space Reddening in the Galaxy"

Henry Norris Russell, "Stellar Masses"

There can rarely have been a more powerful symposium in astronomy, with every talk by an outstanding contributor in each specialist field. It is a matter for regret that the contributions seem never to have been published *in extenso*.

Lest it be imagined that this was a gratuitous exercise in the status quo, one may note that not all of the ideas presented in this list were then widely accepted. Milne championed a cosmological theory that he called "kinematical relativity," which offered an explanation for the observed expansion of the Universe without recourse to Einstein's theory. Instead, he proposed two fundamental time scales that, together, implied a gradual increase in the value of the gravitational "constant." This idea was never very popular and has long since been abandoned in scientific circles. On the other hand, Lindblad's ideas on the origin of the spiral structure found in many galaxies were more or less ignored for many years after, but are now treated with considerable respect.

At the end of the final session, presiding chairman W. H. Wright was given the pleasant task of conveying to the Warner and Swasey authorities the text of a resolution passed by the attendance at the symposium, in appreciation of the company's support for the meeting and for the completion of the telescope. And so the savants dispersed, having baptized the telescope and initiated the formal life of the observatory. They were going back to a world of increasing troubles that would soon shatter true international cooperation in science for most of a decade.

CHAPTER 10

War Years, Struve Years

Putting the New Telescope to Work

McDonald Observatory and—more important—the McDonald telescope were finally open for formal business. Astronomers whose attentions had been diverted by the need to monitor construction projects could turn, albeit briefly, to total devotion to astronomy. The past few months had already been fruitful. When President-elect Rainey declared the observatory open for research on 5 May 1939, Gerard Kuiper had already obtained some 400 unconsecrated spectra of stars with the telescope. Elvey, van Biesbroeck, and Greenstein had had their turns, too. There was nothing wrong in this at all. Indeed, both Struve and the Texas administration would have been upset had the newly finished telescope been left idle. The fetters had been largely symbolic, and the formal dedication removed even them.

A battery of equipment for the 82-inch reflector was completed in short order. The Cassegrain camera and Cassegrain spectrograph had been completed and shipped to Texas in 1938. Moffitt designed the prime focus camera, for which Ross designed the necessary corrector lens. Van Biesbroeck and Keenan designed the mounting for the coudé spectrograph. Horace Babcock, who was transferred from Yerkes to the McDonald staff in April 1940, designed a fast spectrograph for use at the prime focus. Work was undertaken on a whole gamut of lenses, gratings, and other pieces of optical equipment. Eventually, there was an almost bewildering variety of spectrographs and other devices.

In addition, a new 13-inch Cassegrain reflector of 71-inch focal length was installed on the mounting of the unfinished Schmidt telescope, to be used for visual photometry and planetary studies. This was located approximately where the 107-inch telescope now stands.

Another of Babcock's projects was a device called a coronaviser, which used the 82-inch telescope tube as a mounting. A decade earlier, French astronomer Bernard Lyot had invented the coronagraph, with which he could photograph the solar corona without needing an eclipse. An artificial eclipse was created inside the optics by means of an occulting disk and diaphragm, together with scrupulous precautions to minimize the scattered skylight getting into the system. American astronomy was beginning to edge into the electronic age, however, and at Bell Telephone Laboratories, A. M. Skellett decided to replace Lyot's photographic plate with a scanning photoelectric cell. Babcock tested the concept at McDonald, attaching a twenty-foot-long device on the outside of the big telescope. The original design was found wanting, with instrumental noise obscuring the signal from the corona. Rather than chuck the idea, Babcock experimented with modifications to the instrument to overcome its deficiencies, eventually succeeding after about a year's effort.

All was not well with the big telescope itself, however. The images were not good, and Jesse Greenstein was sent to Mount Locke to perform a series of tests that would define the problem. With the problem identified, solutions could be devised. At least, solutions could be devised if the problem were not in the primary mirror itself. For many years, this would be an open question (see Chapter 14).

Struve always maintained, even in later years, that the basic mirror was as good as Plaskett and Lundin had claimed. In his public view, the problem was that the mirror had been tested in the shop, not in the telescope. The difference, he said, was in the mounting. However, Struve's public statements do not entirely square with his more private ones on this score. A few months after the dedication, Elvey, echoing Struve's opinions, wrote to U.T. Comptroller Calhoun that the written version of J. S. Plaskett's address there had contained "a statement relating to the size of the diffraction image that I cannot verify."[89] In simpler terms, the mirror was not as good as its producers claimed. For the moment, however, nothing could be done with the mirror, but something could be done with the mounting.

Superficially, a large mirror seems like an extremely rigid chunk of glass. To the optician concerned with maintenance of its optical surface to millionths of an inch, it seems little better than a rather stiff mass of jelly, capable of deforming under its own weight. When pointed upward, the mirror must be supported uniformly over its back surface by a series of counterweighted axial support pads. To insure its performance when tilted, the mirror must also have radial side supports to preserve the figure and to keep the mirror from flopping about. There must be at least three fixed side supports to define the mirror position exactly, but if they carry the whole load, the entire force will be borne by only one or two of them when the mirror is turned on edge. This will deform the mirror, producing triangular star images. There are additional counterweighted supports to avoid this. It was apparent to Greenstein that there was a serious problem with the support system of the McDonald telescope mirror. Although excellent for spectroscopy, the telescope was much less useful for direct photography. On Greenstein's recommendation, the first of several revisions of the mirror support cell was undertaken. The new system wasn't perfect, but it was better. Nonetheless, van Biesbroeck undertook further improvements in 1944.

Hazards of Astronomy

Working at the Cassegrain focus requires that someone be at the bottom end of the telescope tube at least part of the time. To observe objects close to the horizon, the observer must elevate the observing platform to be able to reach the telescope. If one observes the same object for several hours at a time, then the platform has to be moved during the observation, to keep up with the motion of the telescope offsetting the rotation of the Earth. It is not hard, working at night in the darkened dome, to lose track of how far one is above the floor. Falling off the platform is an occupational hazard for astronomers.

Observing platforms normally are equipped with safety barriers to prevent the astronomer from falling, but they don't always work. Sometimes they don't work because the observer has removed them. This is not a great idea.

Although several people have fallen from the platform of the McDonald 82-inch telescope, including Otto Struve, the distinction of having been the first seems to have been won by Thornton L. Page. On the night of 22 July 1939, Page was using the Cassegrain spectrograph, having removed the safety chain "to make it more convenient to work with the instrument."[90] Page was seated in a chair under the telescope, and every once in a while he had to push the chair back and raise the platform, to compensate for the telescope motion. "I pushed the chair back right over the edge!"[91] He had been some eight feet up in the air, and the fall knocked him out for the rest of the evening. In a letter to Carl Seyfert, Elvey opined that Page had been "terribly lucky."[92]

Legend has it that, shortly after the Page episode, Otto Struve was observing one night when Sergei Gaposchkin came up to the observing floor on a visit. The story of Page's downfall was recounted as a cautionary tale, whereupon Gaposchkin is said to have declared that "I would never fall off of one of these platforms." A few minutes later, he walked off the platform, which was three feet up in the air. There is a very similar story about Gaposchkin in the 1950's, so one does not know where the truth lies here. Maybe he did it twice, or maybe not at all. In any event, Thornton Page was not the last astronomer to walk on air in the 82-inch telescope dome.

Eclipse 1940

An eclipse of the Sun crossed the Big Bend of the Rio Grande on 7 April 1940, occasioning the first major expedition mounted from the McDonald Observatory. Big Bend had only recently become a national park, and a major camp of the Civilian Conservation Corps (CCC) was installed there. Arrangements were made with the National Park Service for the McDonald eclipse team to be based at the CCC camp.

It was considered to be very desirable to communicate results rapidly, not only to the scientific world, but also to the public news media. Consequently, Christian Elvey asked the commanding officer at El Paso's Fort Bliss Army Base if arrangements might be made for urgent radio communications from the eclipse camp to the outside world. The brutal fact of bureaucratic reality appeared in a reply, not from the Army, but from the district adjutant of the CCC (stationed at Fort Bliss): "The CCC Radio Network, which connects the Park with Fort Bliss, may not be used for non-CCC purposes."

Without a radio link, Struve had to be content with a telegram from Elvey (sent from heaven knows where): "Eclipse observed under excellent conditions." A similar telegram was sent to *Sci-*

ence Service Magazine, with an additional note that complete details were being forwarded to the University of Texas Office of Public Relations.

Another solar eclipse occurred six months later, on 1 October 1940. The track of the Sun's shadow cut across the Brazilian jungle. This was not very convenient to McDonald/Yerkes astronomers, and the Sun was not one of the major interests of the observatory anyway, so no expedition was mounted to go watch the eclipse. Nonetheless, this eclipse did provide a serendipitous opportunity for extending the work on the spectrum of the night sky.

During the summer of 1940, the Williams Bay staff was augmented by the temporary presence of Alice Farnsworth, who then taught astronomy at Mount Holyoke College. She and Walter Linke took spectra of the night sky, in collaboration with Elvey's program at Mount Locke. The 1940–1941 school year was apparently a sabbatical year for her, for she had arranged to observe the eclipse from Parabyba, Brazil, and then spend several months in residence at the Córdoba Observatory in Argentina.

One of the things that Elvey had proven in the night sky program was that there were variations in its spectrum with geographic location, particularly latitude. Farnsworth's trip presented him with the opportunity to get southern hemisphere data. A portable spectrograph was lent to Farnsworth, so that she might continue her contributions to the program while south of the equator. Indeed, on her return to Massachusetts, the loan was extended for several months, on condition that data be taken there. The small instrument returned to Mount Locke only when Farnsworth arrived with it, during the summer of 1941. The data that she collected were quite valuable in Elvey's efforts to learn about the composition and astronomical effects of the upper atmosphere.

Clouds on the Horizon

The early months of the McDonald telescope were just as fruitful as had been expected, despite recurrent problems with both the optics and the mechanics. Gerard Kuiper did an enormous amount of work on white dwarf stars. Jesse Greenstein made detailed analyses of the chemical abundances in the peculiar star Upsilon Sagittarii. Franklin Roach (by then at New Mexico State

University), Thornton Page, Daniel Popper, Albrecht Unsöld, and Carl Seyfert made important observational contributions.

The most important discovery during this period—and a major one it was—was made by Pol Swings,[93] a visiting professor from Belgium. He found "forbidden lines" in the spectra of some stars.

Bright lines in the spectrum of a star represent energy emission due to electron jumps in atoms, a change of state from a high energy level to a lower one. "Forbidden" lines are not really forbidden by any physical law. They are simply transitions from or to energy states that are extremely improbable except under very unusual circumstances. That is, they are not forbidden, they only have near-zero probability. They can occur only under conditions of super-low density. Swings' discovery was an indicator that a class of stars called supergiants have atmospheres much more tenuous than those of normal stars. The McDonald telescope was beginning to prove its worth.

There were less agreeable portents of the future. Although the United States was not a formal belligerent in World War II until 7–8 December 1941, Japan and China had been at war since 1937, and the German revanchism first manifested in 1935 flared into open warfare in September 1939, only four months after the McDonald dedication. The threat of war was very much in the air, even during the festivities on Mount Locke. When congressmen feel compelled to stay in Washington rather than join a dedication back home, something serious is afoot.

Astronomy is, by its nature, an international activity, and the effects of approaching war were quickly evident. Albrecht Unsöld, from Kiel, stayed a while in America after the dedication and reached his homeland on the last German liner to cross the Atlantic. He took with him a supply of Greenstein's spectrograms "that kept German astrophysicists in business for years,"[85] according to Greenstein. After Germany's invasion of Poland, the activities of Nazi submarines quickly made transatlantic travel hazardous. Visiting astronomers such as the Belgians Pol Swings and Paul Joseph Ledoux and Norwegian Gunnar Randers were cut off from their homelands for the duration of the war. They and many other staff members of Yerkes and McDonald observatories soon found themselves engaged in quite different pursuits from those of the ordinary peacetime astronomer.

The war brought pressures to bear on the observatory from

many directions, one of which was geopolitical. Today, two generations later, it is difficult even for local people to understand a fear of our southern neighbor. It is almost impossible for those whose families lived distant from the border in that time.

The Republic of Mexico has always been leery of its giant cousin to the north, not without historical reason. During World War I, the Mexicans' fierce independence led to a German attempt to enlist them to invade the United States from the south. There was a diplomatic *cause celèbre* over the "Zimmerman telegram," but nothing ever happened in Texas. Still, the West Texan psyche was severely scarred. Thus, in the early 1940's, there was serious local concern over the remote possibility of an invasion from the south. It came to such a point that Struve contacted the Warner and Swasey Company to inquire about an equipment loan. The equipment in question was that which was used to lift the 82-inch mirror into place. Struve wanted to know if it could be available in case it were deemed necessary to remove the mirror and bury it, to protect it against a military attack. This had been done recently with a large mirror for the Radcliffe Observatory, then being figured in Newcastle, England. The equipment was never lent, and the attack never came, but this is a pungent illustration of the spirit of the times. Keesey Miller recalls that even American citizens with German surnames were required to wear identification badges in that area.

Lest this seem farfetched, consider the experience of some local ranchhands, employees of the Eppenauer ranch. A. R. Eppenauer had come to Texas shortly after serving his German homeland in the First World War. He bought a ranch in the Davis Mountains and had supplied some of the building materials for the observatory. In a remarkable stroke of bad luck, he had chosen an Indian good luck symbol, the swastika, as his brand. The arms pointed in the "right" (i.e., non-Hitlerian) direction, but nice distinctions such as that are lost on most folk. Consequently, when two of his hands drove a pickup to Fort Worth in 1940, they were set upon and beaten and their truck severely damaged by "patriots." Ever since, the Eppenauer brand has been a script EP.

To make matters worse, the Army Air Corps established a bomber training base at Marfa. In a friendly agreement with the State of Texas Department of Parks and Wildlife, the U.S. Army took over Indian Lodge as a habitation for wives of the trainees. During the week, there wasn't much for them to do except talk, and apparently they did a lot of that. Rumors swept around the area like wildfire. One of the rumors was that Immega was distributing German propaganda amongst the "Mexican" workers on the mountain. It was clear to the Army wives, of course, that McDonald was a nest of spies. After all, they had people who spoke with heavy accents and who bore names like Struve and van Biesbroeck. And Immega was not only a strange and unsociable man, but he was also certifiably German, proudly German, irritatingly German. Elvey reported the rumors to Struve.

The director was remarkably unruffled by the charges against his first engineer. He instructed Elvey to tell Immega about the rumors, to say that the observatory was not going to meddle in his personal life and beliefs, but that distributing any political material during work hours was an offense punishable by firing. Nothing concrete was ever brought against him, and those who knew him disbelieve the rumors. Still, Immega soon left the observatory for a better-paying engineering position in New York, after playing nurse and midwife to the machines of the observatory during the first six years of its existence. A decade later, he wrote to Franklin Roach that "it never will leave my system being associated with a bunch of crazy astronomers. [It] is really something to remember."[94]

Christian Elvey's work on the brightness of the night sky attracted attention from an unexpected direction. Theodore Dunham, normally at Mount Wilson Observatory, asked permission to copy one of Elvey's night sky photometers for the National Defense Research Committee. It was wanted for application to a secret British project on "nocturnal devices."

How Can Yerkes/McDonald Survive the World Crisis?

From a vantage of forty years and several cataclysms later, it is difficult to imagine the spirit of the time. Psychologically and industrially, the United States was already at war by late 1940, a year before the attack on Pearl Harbor. The first evidences came when orders for astronomically useful devices, such as prisms from the Bausch and Lomb Company, required special dispensations to compete with national defense needs. It was not long, however, before Struve had to start coping with a more troubling problem: His people had begun to look for ways to serve their country in a time of impending need.

In the summer of 1941, Horace Babcock became the first to announce that he was taking a leave of absence to join the war effort. Effective 1 September, he would work at the MIT Radiation Laboratory, designing military optics. Struve could understand the move, but it was troubling. Babcock had just finished documentation on the prime focus (B) spectrograph, and the coronaviser had only just been brought to a potentially useful state. The man who knew most about both of them was leaving. He would not be the last to leave, either.

The B spectrograph saw considerable service after Babcock's departure, but there is no evidence that the coronaviser was ever used by any of his successors at Mount Locke. A few years after the war, it was dismantled and cannibalized.

Babcock's departure could not have been much of a surprise. Indeed, long before the United States entered the war officially, Otto Struve foresaw a mass exodus of his astronomers, and he began to worry about the survival of the observatory. His first solution was a consolidation of resources, which was submitted in two steps. The first proposal was quite simply to move the giant 40-inch Yerkes refractor from Williams Bay to Mount Locke! As astonishing as this may seem, it was soon matched by an even more audacious scheme: move the entire Yerkes complex to Texas.[95] These ideas never got beyond the stage of internal memoranda within the University of Chicago. Yerkes would stay at Williams Bay.

Rebuffed in his effort to consolidate, Otto Struve rebounded with an even more revolutionary idea, which is now called consortium science. In April 1940, Chicago's President Robert Hutchins forwarded Struve's newest proposal to his opposite number in Austin, covered with the laconic inquiry "What would you think of something like this?" Rainey passed the "Plan for Astronomical Collaboration in Connection with the McDonald Observatory" to Comptroller Calhoun with an equally terse memorandum. The plan proposed that "cooperating institutions should jointly maintain a large observing station to be erected in close contact with the McDonald Observatory."[96]

Struve explained his rationale: "I fear that unless something is done toward equalizing the research opportunities of all astronomers, there will be a gradual deterioration of many observatories which, in the past, have been able to carry on investigations of a quality comparable to that of the largest institutions." He continued, "Let us suppose that a plan of collaboration could be worked out which would be satisfactory to all participating institutions. We should then be able to organize jointly an observing station in the Texas mountains where the McDonald Observatory is located, which would be much more powerful than the present McDonald Observatory alone."

"The plan presupposes," the memorandum continued, "that one of the large foundations will agree to furnish a telescope . . . of the Schmidt type . . . with a spherical mirror of approximately 72 inches and a removable correcting plate of approximately 50 inches in diameter. The instrument is also to be used in the Cassegrain form with a secondary mirror whose shape will correct the spherical aberration of the 72-inch mirror."[96] Nearly half a century later, no larger Schmidt instrument yet exists in the world. According to a news report, it was to be the largest of ten or so new telescopes to be placed on or near Mount Locke. The scale and audacity of the proposal were truly staggering, given the social environment in which they were advanced.

Struve proposed a construction budget of $275,500 and an additional operating budget of $21,500. He computed a charge of $75 per night for time assigned per instrument. "In view of the fact that some of the proposed collaborating institutions have heavy teaching schedules, it is not considered necessary that the astronomers of these institutions go to Texas to carry out their own observations. The plan provides for three observing assistants at each telescope. . . ."

Struve's primary concern was the continued viability of the Yerkes Observatory, but he thought that such a plan held many advantages for other potential participants, too. Struve proposed to invite participation by the Universities of Minnesota, Michigan, and Virginia; Columbia, Princeton, and Indiana Universities; and Swarthmore College. The participating schools would form an observatory council for the "The Cooperative Observatory of American Universities," while McDonald would preserve its own name and independence.

The Yerkes director carried his idea to the astronomical public in an article entitled "Cooperation in Astronomy," in the April 1940 *Scientific Monthly Magazine*. He detailed the advantages of the plan and the superior qualities of the site. Reviewing the progress of telescopic equipment in this country, he remarked, "There has been a general tendency to substitute theoretical studies for observational work. Fortunately, in the United States this process has only started. But in Europe, we have seen the gradual decline of observational work and the rise of theoretical institutions."

Struve was not, of course, opposed to theoretical research. Both Chandrasekhar and Strömgren had been hired on their analytical merits, and Jesse Greenstein emphasizes that the close relation of observation and theory at McDonald/Yerkes during Struve's tenure was an important strength. No, what Struve was saying was "We need more telescopes, not fewer. The only way to get them is to band together."

This was before the days of "big science." Universities were jealous of their autonomies. It is only since World War II that we have seen the establishment of multi-institutional cooperative scientific research facilities operating on a financial scale beyond the capacities of their separate progenitors. Struve's pitch for collaboration fell mostly on deaf ears and blind eyes. It brought a positive response only from Frank K. Edmonson, chairman of the Astronomy Department at Indiana University.

On the Texas side, Rainey and Calhoun were not opposed in principle. They did worry particularly about water supplies, the possible further financial involvement of the University of Texas (already overspent by $100,000), and the preservation of the McDonald identity. Calhoun went so far as to declare that anything astronomical on Mount Locke should be known *only* as the McDonald Observatory, even if the proposed consortium wanted to call itself something else. The matter was tacitly dropped, but the proposal illustrates its author's visionary qualities. In fact, Struve had invented a precursor to AURA (the Association of Universities for Research in Astronomy), which would come into being much later as the vehicle for the establishment and operation of the Kitt Peak National Observatory. Sentiment and world politics were against the realization of such a plan just then.

The idea for a cooperative observatory was stillborn, but it led to an arrangement with Indiana University that endures to the present day. When it became clear that there would be no joint facilities, Struve agreed that McDonald personnel would supply Indiana with "observational material" of Edmondson's specification. Frank Edmondson came to Mount Locke to conduct his own observing run in 1943, and since then Indiana personnel have had an allocation of observing time.

The Yerkes Optical Bureau

Blocked from creating a national observatory, Struve knew moments of discouragement. After the June 1941 invasion of his native Russia by the Nazis, Struve threatened to join the U.S. military himself. Whether this was a real thought or a power play is unknown, but subsequent events admit either interpretation.

Struve was faced more critically than ever with the problem of keeping his organization functioning in the face of a war that had requirements for his people. He had no lack of patriotic feeling, but he was convinced that the observatory had to continue to operate to avoid disintegration. He wanted to make sure that there would still be an observatory *after* the war. The way to do that was to find ways for the Yerkes/McDonald staff to contribute to the war effort without leaving.

That is the background to this curious passage from the Annual Report for 1941–1942: "Because it was recognized that most astronomers could at best function only as second-rate physicists if transferred to physical war research laboratories, an optical bureau was organized at Yerkes with Louis G. Henyey, van Biesbroeck, and Greenstein as principal workers. Towards the end of the year this bureau was engaged in war work of considerable magnitude."[97]

The "second-rate physicist" remark is a red herring, given the lie by numerous examples. Following Babcock, Christian Elvey converted very successfully to rocket research at the Naval Ordnance Test Station at China Lake, California, where he eventually became research director. Gerard Kuiper was involved in a number of military projects.[98] John O'Keefe worked for many years for the Army Corps of Engineers. The Yerkes/McDonald list is longer, but in fact astronomers from observatories all over the country contributed in many diverse ways to the technical progress that had such a major influence on the course of World War II. Otto Struve was clearly torn between furthering the war effort and preserving the McDonald organization, which he regarded as an essential part of the western culture in defense of which the war was being fought. The Yerkes Optical Bureau was Struve's means of keeping some of the astronomers at home.

His first efforts to obtain government contracts for the Optical Bureau were launched months before the formal entry of the United States into the war. They met with no more success than had his cooperative observatory. David DeVorkin, historian at the Smithsonian Institution National Air and Space Museum, in a recent study[99] of the Optical Bureau suggests that the initial proposals were somewhat inept. Escalation of the national emergency in the aftermath of the attack on Pearl Harbor gave officials second thoughts, however. Military optical work was done at Williams Bay throughout the war, with Greenstein and Henyey devoting essentially full time to it, occasionally observing as well. Struve could breathe easier about his staff and postwar problems.

Coping with the War

That is not to say that the astronomical business went along as usual. Once the United States entered the war, many of the Yerkes/McDonald staff resigned or secured leave to take up arms or to do defense research elsewhere. These included Page, Elvey, Popper, Chandrasekhar, Swings, Harold Weaver, William P. Bidelman, and Kuiper. Most of them came back after demobilization, but their departures left a great hole in the activities of the observatory.

The departure of Christian Elvey in 1942 was a particularly serious blow. Elvey had been "acting for the Director" for seven years, eventually being granted the title of Assistant to the Director. He was both a good astronomer and a good manager. It was impossible to replace him, given the circumstances.

Dr. Elmer Dershem, approaching retirement in the University of Chicago Physics Department, agreed to go to Mount Locke for the duration of the war. Dershem was a first-class instrument man and knew his way around a machine shop. His primary task was to keep the telescope and its instruments in working condition until normal times returned. He was explicitly *not* authorized to act for the director. Yerkes astronomers would take multi-month duty tours at Mount Locke, and the ranking astronomer on the mountain was temporarily in charge. The first of these was the stranded Belgian, Pol Swings.

Measures to cope with restrictions and to encourage conservation increased as the war went on. A commissariat for the sale of canned goods to the staff was opened in the dome. Pocket billiards became a favorite pastime, with food from the commissariat as stakes. Then as now, the support staff were noticeably more skilled at the game than the astronomers, and their families were grateful for the presence of the pool table.

A business office was established in Fort Davis as a rubber and gasoline conservation measure, so that truck trips between town and mountain could be reduced to once in five days. Observing runs with the main telescope were long and rugged experiences. On one occasion, Swings lost twenty pounds in weight during a month at the mountain.

Part of the problem was that the observatory was understaffed due to war manpower needs. It was not only the astronomers who were called away; in 1944 alone, four of the experienced support staff were drafted into the military. (Ironically, since they were from a semi-desert, semi-mountainous area, they all were sent into the U.S. Navy.) The observatory was able to continue functioning primarily because Otto Struve and University of Texas Vice-President J. Alton Burdine persuaded the Fort Davis draft board to grant an occupational deferment to José (Joe) Rodríguez, while Tommy Hartnett was deferred for physical reasons. In seeking Rodríguez' deferment, Struve argued that the continued operation of the observatory was indirectly vital to the war effort. "All of the astronomers are engaged in important war research," he said. This work was not undertaken on Mount Locke, but could not be carried out without the observational facilities that the working observatory provided.

The workers who remained could not stay on duty throughout the night to maintain the diesel power house that provided electricity to the mountain. Night assistants were sometimes laborers untrained in astronomy, such as Hartnett. When they were available, they would usually work an eight-hour shift and then go home to sleep. After that, the astronomers had to be their own night assistants. For his own observing sessions, Struve got around this problem by teaching Hartnett how to operate the telescope. On long, cold winter nights, they would alternate, one hour on, one hour off. Every other hour, they could each get their hands warm and rest or, in Struve's case, write.

Hartnett also assisted George van Biesbroeck in his observations at prime focus, and the descriptions are hair-raising. The Belgian was about a foot shorter than average, and he had trouble reaching the camera attached to the upper end of the telescope. Access is from pulpits swiveling on a curved staircase slung high on the inside of the dome. The pulpits are equipped with safety railings, to minimize the turnover rate among observers. The barriers kept the diminutive astronomer from the camera, so he would stand on the railing 45 feet above the floor and lean out, catching himself on the telescope—in near-total darkness. "He would scare me to death sometimes," Hartnett recalls. "It's dark, and he's standing on the rail, looking down the telescope and telling me which way to move it. One night, it was 7 degrees below zero."[100] Van B lived far beyond his threescore and ten, despite tempting fate at the prime focus of the McDonald telescope. He was more than threescore when the management finally ordered him to stop observing in that precarious location.

The operating engineer Arch Garner played a crucial role in maintaining any kind of operation early on. The dome and telescope contained some thirty electric motors, whose functioning relied upon huge banks of mechanical relays. Not only did they make a lot of noise, both mechanical and electrical, but they were not always perfectly reliable. Jesse Greenstein recalls how Garner would come out in the middle of the night in answer to an astronomer's distress call. Clad in pajamas, he would climb a ladder to poke at a balky relay with a broom handle until it worked again. Garner left the observatory midway in the war, being replaced briefly by Fred Belland from Yerkes and eventually by Joe Rodríguez.

What is surprising is the amount of astronomical research accomplished during these years at McDonald despite depletion of staff and shortages of every kind. The six years 1939–1945 saw 106 numbers of the *McDonald Observatory Contributions* published. Of these, Struve and Swings accounted for 55. With nearly everyone else occupied with war research, the old team of Otto Struve and George van Biesbroeck provided the core of the observing staff from 1943 on. They were supported by W. A. Hiltner, Armin Deutsch, W. W. Morgan, Daniel Popper, and of course Greenstein and Henyey. In addition, Carlos Cesco and Jorge Sahade came from neutral Argentina and Guido Münch from neutral Mexico to participate in the work of the observatory. The main thing was that the telescopes were running. A fundamental astronomical principle was being observed: An observation not taken is lost forever, so when the sky is clear, take data. Worry about what to do with them later, when the sky is cloudy or when you have more staff.

There was an occasional striking discovery during this period. Gerard Kuiper, on a brief break from his government duties, found the first known atmosphere of a satellite when he detected methane on Titan, the largest moon of Saturn. In general, however, the observational programs were geared to accumulating data for longer-range applications. In keeping with this, a discussion of the astronomy accomplished at McDonald Observatory during this period will be deferred to the next chapter.

War, as with many human activities, can have far-reaching and unexpected side effects. The end of World War II brought many of the departed astronomers back home to Yerkes/McDonald. Perhaps Otto Struve would not have been so pleased at this development had he realized that "you can't step into the same river twice." The war changed the rules of the game, and its end set in motion a process that would lead to Struve's departure.

CHAPTER 11

Triumphs and Transitions

At the end of the war, astronomers began to be released from defense work and returned to their universities and observatories. McDonald ended the war with most of its advantages intact. The organization had managed to carry on, despite the stringencies. The greater obstacles encountered by many other astronomical research institutions contributed to making McDonald the world's outstanding center of stellar research. Mount Locke provided superb observing conditions; Yerkes and the University of Chicago furnished supplementary equipment and were home to a galaxy of astronomical talent. These included not only excellent observers, but also some of the most profound theoreticians in the world. An additional advantage was the production of *The Astrophysical Journal* on the spot.

When the war ended, the cream of the astronomical world flocked to Chicago and McDonald. Young people whose careers had been frustrated for several years rejoiced to be free once more from the obligations and dangers of a war-ridden world. The liberated astronomers of Europe had heard of the important advances made during the lost years and hastened to try to catch up. Even some astronomers from the Axis countries tried to come, but the U.S. government frowned upon their presumed loyalties and denied them admittance.

The roster of those who came is too long to enumerate here, and may be summarized simply by saying that anyone who was anybody in world astronomy at that time did his best to visit this mecca of astronomical science. Those who succeeded in doing so responded to the intellectual challenge and, in the clear skies and wide open spaces of Jeff Davis County, combined intensive study with picnics and barbecues. In the months and years just after the war, Mount Locke was a cosmopolitan place to be—cosmopolitan in terms of the company one could keep, if not the surroundings. Nearly every observational astronomer in the country spent some time there, and there was a glittering array of foreign visitors.

Cosmopolitan, perhaps, but hardly easy. Just getting up the hill was often a hazardous task. The poor quality of postwar gasoline assured that vehicles going up the 17 percent grade often stalled in midroute. Standard procedure was to chock the wheels with large stones, wait until the car or truck would again conquer gravity, and drive off—leaving the rocks behind. This practice provided an interesting slalom event for subsequent drivers.

There were good times, though. In good weather there were picnics to a place called The Rockpile, a spot where ancient geological forces have left a pyramid of smooth boulders several hundred feet across and some fifty feet high. These outings were considered big social events, welcome respites from the austerity of the mountain. Mary Struve was apparently never comfortable with the informality, so the Struves rarely participated in the group picnics.

The flood of postwar visitors included several astronomers and at least one engineer from the Soviet Union. A large delegation spent several months in the United States, touring all of the major astronomical facilities. They split into several groups to increase their coverage. Those who came to McDonald at one time or another included Aleksandr Aleksandrovich Mikhailov, Grigori Abramovich Shajn and his wife, Belageya Fedorovna Shajn, and four others. Several of them stayed long enough to observe with the McDonald telescope, and some of them even came back a second time. They came more to observe the telescope than to observe the sky. On their return home, the Soviet government asked the Warner and Swasey Company for a bid on constructing a duplicate of the 82-inch telescope. The company was not enthusiastic about the idea in the abstract, and the thought of building such a fine instrument for a *communist* government was really abhorrent. Warner and Swasey's response was in the finest capitalist tradition: they overpriced it by a factor of four, assuring that the matter would be dropped.

Several of the early inhabitants of Mount Locke, especially the Kuipers, Pol Swings, and the van Biesbroecks, were socially active with the citizenry of Fort Davis. According to "Bit" and Lucy Miller, they often participated in potluck suppers either down in town or on the mountain. In a reference to a then popular comic strip, these were known locally as "duck dinners"—come to din-

ner and bring the duck. Dinner was often followed by parlor
games, usually charades. On one such occasion, Lucy Miller had
an inspiration and seated Gerard Kuiper on a kitchen stool with a
cap on his head. Unfortunately, there was little mystery; he was
quickly recognized as a charade for Dutch Boy Paints!

The observatory served—as it still does—as a social leaven in
the life of the region. Although many of the ranchers wanted to
keep the area free from outsiders, they had welcomed the establishment of the observatory out of sheer goodwill and local pa-
triotism. From the beginning, it has provided both a field for local
employment and an avenue for upward economic mobility and
further education.

There was an uglier side to this idyllic picture. Neither West
Texas, the University of Texas, nor the Yerkes administration yet
aspired to those levels of fairness now considered obligatory in
American life. Most of the manual work was done by a fluctuating
crew of Spanish-speaking Americans who were usually referred
to as "Mexicans," although many of their families had been Texas
residents and citizens for several generations, and several of
them had been quite American enough to have served in the U.S.
armed forces during the war. Indeed, some were descended from
former Indian fighters who settled near Fort Davis after being
mustered out, and thus bore north-European rather than Spanish
surnames: names such as Hartnett, Dutchover, and Webster.
Working conditions were not very good, and these "Mexicans"
were often accused of shiftlessness, absenteeism, and drunken-
ness. Perhaps some of the accusations were valid, but more often
they seem to have been reflex actions of unconscious bigots; the
letters of the temporary superintendent, Elmer Dershem, are
particularly uncomfortable to a present-day reader. Yet among
this group of "Mexicans" were outstanding men who gave the observatory competent and dedicated service over many years. In
particular, Tommy Hartnett did so throughout nearly the whole
of his working life.

Postwar Plant Improvements

Harnett participated in one of the major improvements of this period, when the University of Pennsylvania lent a telescope to Mc-
Donald. In August 1949, he and Paul Jose (of whom more later)
furbished up the old 1½-ton pickup truck and drove first to

Yerkes and then to the Cook Observatory in Philadelphia. The Cook Observatory was the property of the university, but located on the grounds of a private (and expensive) home. Constrained on pain of dire penalties not to damage the front lawn, they loaded the 10.25-inch refractor and brought it to McDonald. This remarkable instrument had a field measuring 20 × 24 inches. It was to be used for asteroid and Milky Way photography. It was accompanied by two Ross-Fecker cameras of 4 inches aperture, and it was installed in a rectangular building with a rollback roof on the south side of the mountain. The scheme was to photograph the ecliptic in the region opposite to the Sun, where asteroids would execute their retrograde motions against the star background, as the Earth overtook them in its faster orbital motion. The moving objects could be identified by their change in position from night to night, and their orbital parameters thus deduced. The Cook telescope had a steady output of some hundred plates per month until April 1956. Although the instrument has since been returned to Philadelphia, the building is still there and houses a patrol camera occasionally used for photography of comets.

W. A. Hiltner was responsible for many instrumental improvements during this period. In 1945, he introduced photomultiplier tubes into stellar photometry. A year later, he built a photometer incorporating a 1P21 RCA tube of the kind one can still find in use today. In 1946, a grant of $12,000 from the University of Texas permitted him to begin construction of a new coudé spectrograph which is, with few changes, the one in use at the 82-inch telescope today. In 1948, he and Thornton Page oversaw the final connection to the external alternating current power grid, and it was no longer necessary for anyone to babysit the diesel generators throughout the nights. Hiltner also installed a new drive for the 82-inch telescope and used photoelectric photometers recording on Brown-type chart recorders in his studies of stellar magnitudes. Meanwhile, Page perfected the prime focus B-spectrograph, which he used in his observations of the kinematics of multiple galaxies. This was destined to become the major instrument in the long series of measurements of rotation curves of galaxies to be made by Margaret and Geoffrey Burbidge in the next decade.

Gerard Kuiper introduced a major innovation when he mounted an image tube at the Cassegrain focus for planetary observations.

In view of the important role to be played by such devices in future astronomical applications, a word of explanation is necessary. An image tube is a light amplification device, a sort of cousin to both a television tube and a photomultiplier. The light is collected in the same way as for a photograph. Instead of meeting a photosensitive emulsion, however, it falls on the front surface of a tube whose coating emits electrons into the interior of the tube when struck by light photons. These electrons are accelerated in the vacuum tube by electrical or magnetic fields, applied so as to reform the image on a television-like phosphor surface. This secondary image is stronger than the original would have been, and it forms the observational record. It is thus a means of making a telescope perform as though it were larger than it really is. The general principle has, since Kuiper's time, been realized in many ways, using phosphor screens, photographic plates, and solid-state electronic devices.

Not all of the instrumental developments turned out well. The 20-inch Schmidt optics were finally finished in 1946, after the hiatus caused by the war. The mechanical design of the device was faulty, however, and the mirror was chipped during the unsatisfactory testing. Kuiper reported to Struve that the mirror was put into storage and the holes in the tube stuffed to keep the insects out. Nothing is ever said again about the 20-inch optics. A 28-inch Schmidt was considered briefly, but that idea too was dropped.

Struve had one last fling with his dream of a multi-university collaborative observatory at Mount Locke. Chicago, Texas, and Indiana were still eager, and astronomer Willem Luyten convinced the president of the University of Minnesota to throw a hat into the ring. The giant Schmidt had by now been abandoned, possibly due to the troubles that had been encountered with a much smaller one. The plan presented to the Rockefeller Foundation in 1947 called for constructing a 100-inch reflector "of a special design." Eventually, Struve concluded sadly that the powers at the foundation thought that they had spent enough on astronomy, with the 200-inch at Mount Palomar still under construction. Once again, the plan died, this time for good.

New Faces in the Dome, New Hands on the Controls

George van Biesbroeck passed his sixty-fifth birthday in 1945, and University of Chicago regulations mandated that he retire.

Although in his case, "retirement" became a euphemism for nearly thirty more years of strenuous continued activity in a variety of astronomical and instrumental fields, to Struve this must still have meant the loss of a valued ally and friend in departmental politics. The little man had been a happy co-conspirator in Struve's Texas ambitions from the first, even before the McDonald bequest and the interuniversity agreement. He had been a pioneer on the mountain and was immensely popular among the Fort Davis residents, and marvelous with their children. His energy and enthusiasm were legendary. Even in extreme old age he retained his panache, observing with the 82-inch telescope on his eighty-second birthday. In a similar circumstance at the 84-inch instrument at Tucson two years later, he remarked that he could hardly bear to wait for the 150-inch reflector then under construction. He did make it to the 90-inch Steward Observatory telescope, six years after that. His retirement barely affected his astronomy, but the psychological effect must have weighed on Struve in the next few years.

With the war over, it was time to replace Dershem with a resident astronomer, someone whose function would be more than simply to keep the machines running. In the fall of 1945, Struve invited several people to apply, including Paul D. Jose (rhymes with rose, not the Spanish name José). In 1939, Jose had been a public school teacher in New Mexico, but he was a trained professional astronomer. He had attended the dedication of the McDonald telescope, but was unknown to most of the astronomers, so he was identified in the group picture published in *Popular Astronomy* as an anonymous "newspaperman." By 1945, he had become assistant director of the Steward Observatory of the University of Arizona. Interested and qualified though he was, Jose failed to get the nod. With three young children, Jose posed the condition that the observatory be responsible for their transport to and from school in Fort Davis. The University of Chicago refused, although Struve commented at the time that this would ultimately be a necessary responsibility. Instead, Arthur Adel, of the University of Michigan, was appointed to the position.

In due time, Adel arrived at Mount Locke, accompanied by his wife. The next morning, the new resident told Kuiper that he was leaving, giving the apparently unintended impression that he was going into town for supplies. Not so! Mrs. Adel had taken one look at the isolation of McDonald and declared that she was on no account going to live in such a place. Kuiper tried to talk Adel out of the decision, but to no avail. Together, they went to Hiltner, who had recently become assistant director. More talk brought the same result. The Adels were leaving. Off they went for good, back to the Physics Department at Ann Arbor, Michigan. They were on the mountain less than twenty-four hours.

The search for a resident astronomer was reopened. Kuiper suggested that the pioneer of radio astronomy, Grote Reber, might be available and interested. He was in the Electrical Engineering Department at Wisconsin, but had discovered that astronomers valued his work much more than did his engineering colleagues. Struve demurred and reopened negotiations with Jose, by now on the mathematics faculty at Washington and Jefferson University (Washington, Pennsylvania). Jose decided that he wanted the job badly enough that he agreed to solve the school problem by maintaining two residences, one on the mountain and the other in town. McDonald had a new resident astronomer. It illustrates the climate of the times that Struve felt obliged to reassure the Texas regents that Jose was from good Pennsylvania stock, rather than from below the Río Grande.

Things were not going well back in Williams Bay and Chicago, however. The bright young subordinates of 1940 had come out of the war as the young Turks of 1946. They knew that they were very good, and they knew about the technological and social revolutions that the war had set in train. The autocratic methods that Struve had learned in his upper-crust Russian family fit poorly into the new vision of the world. Disagreements began to develop over observatory policy and planning for the future.

There is evidence, but not proof, that Struve's reaction to this development was to fall back on the levers that he had so carefully built into the Texas-Chicago agreement. According to this document, his resignation would abrogate the agreement, if the two universities could not agree on a successor. According to Thornton Page, he offered his resignation several times as a means of winning policy disputes either with his staff or with the University of Chicago administration.

Another in a series of crises erupted when Chandrasekhar was promoted from full professor to "distinguished service professor" in the University of Chicago. He and Kuiper had arrived at Yerkes together and had been promoted in parallel up to that point. The new recognition of Chandrasekhar's genius, in his own words, "created an imbalance that troubled Struve."[101] Struve hit upon a plan to keep the peace. Kuiper would become director of the Yerkes and McDonald observatories. Struve would retain the chairmanship of the Astronomy Department, and he would take on the new position of "honorary director" of the two observatories.

In fact, Struve was giving up nothing. He very explicitly notified the University of Texas administrators that he was still in charge. Kuiper had no authority to act without Struve's approval. There was no need to reconsider the Texas-Chicago agreement.

On 1 July 1947, Gerard P. Kuiper became the second director of the McDonald Observatory, with Hiltner as his assistant director. To the outside world, the mantle had passed, but in reality the change was only cosmetic. Struve is said to have deeply regretted these arrangements later. In any case, neither side of the configuration lasted very long.

At the beginning of Kuiper's first directorate, several of the better-known researchers moved on to new posts, where appointments were again being made after the war's interruptions. These included Greenstein, the Dane Kaj Strand, and Dutch Hendrik van de Hulst. They were replaced by a new pool of younger talent whose names are now well known: Nancy Roman, John Phillips, Arthur Code, Arne Slettebak, Marshall Wrubel, and Robert Hardie. Page had returned, as had Swings, while Strömgren and Mogens Rudkjøbing were there from Denmark, as well as J. H. Oort and Adrian Blaauw from Holland.

Kuiper's first tour as director was less than a complete success, even given the fact that Struve held the ultimate authority. While he was a superb astronomer, Kuiper was apparently unable to delegate authority for even minute details of McDonald administration. He was deeply concerned with the color of the concrete used in the construction of paths. When one of the houses needed paint, he was seen mixing paint to get exactly the right color. Meanwhile, the day-to-day functioning of the observatory was haphazardly attended. There was no money to purchase proper weights for the observatory's fundamental sidereal timekeeper, a pendulum clock; paper clips and thumbtacks would do. Eventually, Jose had a blazing row with him in the parking lot.

The resident astronomer at Mount Locke was not the only person to have problems with Kuiper's directorship. Most of that story, however, pertains more to the University of Chicago than to McDonald and so is not relevant here. The surrogate directorship, with Struve in the shadows, didn't work. Kuiper asked to be relieved of his administrative duties as of the close of the 1948–1949 academic year. Scientific matters were passed to an Observatory Council, under Struve's chairmanship. According to Chandrasekhar, this was yet another smokescreen behind which Struve exercised power. After a few months, he no longer bothered to consult the council on important matters.

Kuiper was not the only one worn down by the daily grind. Jose left after only three strife-filled years. Struve himself decided to get out. Having decided to leave Chicago, he resigned the chairmanship of the department, while hanging onto his honorary directorship. Whatever the titles, he was the de facto director of the Yerkes and McDonald observatories until the day he left Chicago. Administrative responsibility fell upon Subrahmanyan Chandrasekhar, who became acting chairman of the Astronomy Department on 1 January 1950, succeeding Struve until a permanent replacement could be found. This marked the first time that the functions of director and chairman had been truly separated.

In spite of the fact that the Yerkes-Chicago astronomical organization was afflicted with sometimes bitter differences, many of those who left for other institutions were designated "research associates" and continued to collaborate with their former colleagues and to observe at both Williams Bay and Mount Locke.

On 1 July 1950, Struve took up appointment as professor and chairman of the Astronomy Department at the University of California at Berkeley, thus severing his connection with the organization that had brought him out of the slums of Istanbul, and which in return he had brought to the pinnacle of astronomical endeavor. For the next six months, McDonald Observatory would have no director, honorary or otherwise.

Struve's Scientific Legacy

Officially, Otto Struve gave up the Yerkes/McDonald directorship in 1947, but he was evidently still the predominant force in observatory affairs until his departure from the University of Chicago in 1950. The science performed during these three years was as much a product of Struve's vision and planning as were the accomplishments during his administration. Consequently, in assessing "the Struve Years" of the post-dedication McDonald Observatory, we shall cover the period 1939–1950.

Otto Struve has been described as a modest man, with a stern devotion to his science. As a spectroscopist, he had spent many years repeatedly looking at a spectrum plate through a microscope with one eye and at the position of a measuring head with the other. This severely damaged the convergence of his eyes. As a result, he sometimes presented a bizarre appearance, when one eye would wander off in direction rather in the manner of a chameleon.

Struve was a reserved individual in dealing with his staff, indeed with the entire world. Close associates have said that they never saw the inside of his residence at McDonald. Thornton Page recalls that Struve's mother was the only person who used his given name. To everyone else, including his wife at least in public, he was Mr. Struve. The only known exception was the only time that Page ever saw him thoroughly relaxed: Page had lured him off to a St. Patrick's Day party in a Chicago night club, during which time the man in the green hat was called Mr. O'Struve!

At work, he was a strict disciplinarian. He did not shrink from dismissing subordinates who did not meet his own high standards of achievement. He expressed his administrative displeasure when Strömgren and Chandrasekhar exchanged the Yerkes offices that he had assigned to them. On the other hand, his instructions for duty schedules at Mount Locke state: "It is planned to operate the big telescope during all clear hours of the night. . . . Astronomers must be on duty during all hours of the night, whether the sky is clear or cloudy, in order to start work at once if it should clear up suddenly. . . . The normal observing load should be five hours every other night. . . . Accordingly there must be a minimum of four astronomers on Mount Locke at all times. . . ." This is a lighter load than is presently shouldered by nonresident astronomers, but they now come for relatively short periods, rarely exceeding a week. Struve's precepts represent a heavy enough load for resident staff or transients required to spend several weeks at a time on the mountain, with long hours of daytime preparation and analysis in addition to the hours on the telescope.

Struve directed the observatory under adverse conditions. First, there were the difficulties attendant on bringing a complex organization into operation on a remote site. Once these problems were surmounted, there followed the stringencies and disruptions of wartime and the frustrations of the slow return to a normal world. Paradoxically, his outstanding record of achievement was aided by the fact that there was very little astronomical activity elsewhere in the world during the early 1940's. There were important exceptions, of course, perhaps the most oustanding of which was the discovery of Walter Baade at Mount Wilson Observatory that there were two distinct populations of stars, the old and the young, a discovery made possible by the wartime blackouts in Los Angeles.

In any event, under Otto Struve's guidance and influence, the Yerkes-McDonald collaboration laid a large proportion of the foundations of modern astronomy, mostly with observations from the McDonald telescope. Lest this assessment of the importance of McDonald seem too partisan, judge the esteem of the astronomical world for the achievements of the Yerkes-McDonald-Chicago group in this era by its remarkable collection of honors.

First recognition came in 1944, when the British Royal Astronomical Society conferred its Gold Medal on Struve. Then, in a five-year period around 1950, came a deluge of honors for the Struve team. Struve himself was elected vice-president of the International Astronomical Union and would later be its president. He was elected president of the American Astronomical Society. The Royal Astronomical Society invited him to London to deliver the prestigious George Darwin Lecture. He was awarded the Draper Medal of the U.S. National Academy of Sciences and the Janssen Medal of the Société Astronomique de France.

Gerard Kuiper received his own Janssen Medal, as well as the Rittenhouse Medal of the Franklin Institute and election to the National Academy of Sciences. Chandrasekhar won the Adams Prize of Cambridge University, the Russell Prize of the American Astronomical Society, and the Gold Medals of both the Royal Astronomical Society and the Astronomical Society of the Pacific. Aden Meinel won the Adolph Lomb Medal of the Optical Society of America. All told, this is a record of concentrated awards probably unequaled in the history of modern astronomy.

Struve considered Williams Bay and Mount Locke to be a single observatory. Consequently, it must be understood that the litany that follows is a composite of work accomplished at the two sites, since it is often impossible to unravel them. Nonetheless, the McDonald telescope was the principal jewel in the Yerkes/McDonald crown. The legacy of Otto Struve is almost indistinguishable from the history of McDonald Observatory up to 1950. The astronomical research done prior to the dedication of the McDonald telescope has already been discussed in some detail, so the following is restricted to the period 1940–1950.

During the Struve years, Jesse Greenstein developed methods of chemical analysis of stellar atmospheres from the study of their spectra. Although his original methods have been greatly refined by later workers, his results have stood the test of time.

Thornton Page's observations of planetary nebulae were interrupted by the war, but he returned to observe pairs of galaxies affecting each other by their mutual gravitation. He discovered the extremely high ratios of mass to luminosity required to explain their orbital velocities, an early omen of the current problem of "missing" (i.e., invisible) mass in the Universe.

Daniel Popper concentrated on binary stars by making radial velocity observations for Indiana University. This is virtually the only way to determine stellar masses without making critical assumptions about the way in which the Universe is built.

Philip Keenan concentrated on red giant variable stars, while collaborating with W. W. Morgan on an atlas of stellar spectral types and luminosities that would set the standards in this field for a long time to come. Canadian Gerard Herzberg, later a Nobel laureate, made infrared observations of stars and turned his attention to a topic still of prime interest today, namely the isotope ratio C^{13}/C^{12}, which is determined by evolutionary effects on the structure of stars. Many of these studies were aimed at exploring the observational implications of Chandrasekhar's new theory of stellar evolution.

Throughout this time, Gerard Kuiper pursued diverse researches distinguished by a single master theme: an interest in multiple astronomical objects. This proliferated into a great variety of important derivative investigations, both observational and theoretical.

Kuiper had a special interest in faint nearby stars, such as M-type (cool red) dwarfs and white dwarfs. These may be very numerous in space, but are hard to pick out because they are so faint. Kuiper discovered many such objects and analyzed the stellar population in the neighborhood of the Sun. One of his major conclusions was that at least half of all "stars" are really multiple systems of two or more stars orbiting one another. This means that at least two-thirds of all stars are members of multiple systems, if the sample of nearby stars is typical of the general population.

Kuiper determined that the class of so-called "sub-dwarf" stars, objects often less luminous than ordinary dwarf stars of the same color, have very high velocities with respect to the Sun and are distinguished by very low abundances of the heavier elements. This proved to be a pointer to the subject of stellar evolution and galactic structure, as it would develop in the next decades.

A request to review a book on the origin of the solar system turned Kuiper's attention to what would be a lifelong study, the origin and physics of the multiple systems represented by stars possessed of planets, beginning naturally enough with our own solar system. Toward the end of the war he made spectrographic studies of the planets and their major satellites, including the discovery of the atmosphere of Titan, already mentioned. In the course of his war work, he learned of new infrared detectors developed by both sides. When the American detectors were declassified in 1946, he soon applied them to his planetary studies. He became one of the first astronomers to consider chemistry as giving clues to planetary origin and evolution. This quickly resulted in the discovery of carbon dioxide in the atmosphere of Mars, a discovery that was announced (among other places) in the illuminated newsstrip atop the Boston Travelers Building.

Among Kuiper's more popularly appreciated discoveries were those of Miranda, the fifth satellite of Uranus, in February 1948, and of Nereid, the second satellite of Neptune, in May of the following year. The latter of these occasioned a satirical contribution to a Chicago newspaper:[102]

HEAVENLY BODIES

Kuiper discovered Moon #30,
three billion miles away.
It is to mankind no great boon,
this finding of another moon,
A satellite that only he,
the great astronomer, can see.
Heavenly body! Fine, we say,
but not three billion miles away.

Carl S. Junge

Kuiper did intensive observational work on measuring the angular dimensions of satellites, using a device invented by Henri Camichel at the Pic du Midi Observatory in the French Pyrenees. A disk of known size was subjected to the same enlargement effects caused by seeing and irradiation as affect actual celestial objects. Comparison of the disk image with that of a planet or satellite permits an estimate of the real size of the body. Kuiper measured many solar system objects in this way.

He also began an intensive study of the features of the lunar surface, seeking to read the history of their formation, using the standard archaeological maxim that overlying material is later material. In the course of this work, he used large numbers of lunar photographs taken at McDonald and elsewhere. These were projected optically onto the surface of a sphere and rephotographed from a different angle chosen to produce the effect of a picture taken from directly above the surface, rather than from Earth. The results, as near to an undistorted map of the lunar surface as possible from purely Earth-based material, were eventually published as the *Photographic Lunar Atlas*. Massive though these efforts were, from 1946 Kuiper was simultaneously engaged in the editorship of his first astronomical compendium, *The Solar System*, published in four volumes during the decade 1953–1963. Whatever his merits and/or frustrations as Director, Gerard Kuiper was leaving everyone else in the shade as an observational astronomer.

During all this time, van Biesbroeck observed regularly at McDonald. His large numbers of double star observations are now the basis of a great many double star orbits. He sometimes worked with visitors such as Swings on the spectra of comets. He observed solar system objects, such as Kuiper's new satellite of Neptune, Nereid, to determine their orbital parameters. Stories of van Biesbroeck are legion. Despite official retirement, he remained as active as ever, going on solar eclipse expeditions to such places as Brazil and the Sudan. He even advised the Belgian government on the establishment of an observatory on the equator in the then Belgian Congo. This telescope was in fact built, never installed, and almost destroyed as scrap. One of the authors of this history had a hand in alerting the government of Congo-Kinshasa to the fact that the nation possessed a telescope with a nearly-horizontal axis. The mirror is now an exhibit at the Royal Greenwich Observatory in England.

Jan Oort and W. A. Hiltner broke new ground by studying the distribution of intensity and color in extragalactic nebulae using a photoelectric photometer. This led on to studies of stellar populations, the evolution of galactic structure, and young stars in galactic spiral arms. In California, Allan Sandage would carry the ball further and lay the observational foundations of the theory of stellar evolution.

In July 1948, Hiltner discovered that radiation from distant stars was often polarized by the interstellar material through which it had passed on its journey to the Earth. This phenomenon is due to the shape or alignment of the dust in the interstellar medium. As a result of this discovery, polarization became a significant tool for studying the space between the stars. This is not the only cause of polarization, as Hiltner himself was to find nearly a decade later. In the Crab nebula supernova remnant (and some other objects), the polarization is due to intrinsic physical conditions.

During this immediate postwar period, Struve demonstrated that he was just as ready to engage new technologies as he had been a decade earlier. Then, it was Pyrex mirrors and vacuum deposition. Now, it was the first timid steps toward space astronomy. The U.S. Navy had acquired a number of the German V-2 rockets (the Germans called them A-4) and was testing them at White Sands Proving Ground, New Mexico. That was close enough that some of the launch trajectories could be seen from Mount Locke. The Yerkes astronomers were not mere spectators, though. Some of those rockets carried Yerkes spectrographs. This may have been the first attempt at doing astronomy from outside Earth's atmosphere.

Struve's own research concentrated on bright peculiar stars, often in collaboration with others. Nearly every star that Struve chose for study now ranks as the prototype for an entirely new branch of stellar astronomy.

Pleione, one of the Pleiades, set the tone for the subject of tenuous gas shells surrounding stars. Studies of α^2 Canum Venaticorum, an A-type metallic star with rare earth metals, prefigured subsequent studies by the Burbidges and others. The stars β Cephei and β Canis Majoris are both small-amplitude pulsators, prototypes of hot stars pulsating with more than one period simultaneously. T Coronae Borealis is the type star of a kind of recurrent nova, a star that undergoes repeated explosions without

destroying itself. Struve also studied the spectra of the longer period pulsators (the Cepheid variables), and with Hiltner looked at
RS Canum Venaticorum, now the prototype of a group of variable
stars having a mottled or spotted surface.

A collaboration between Struve and George Herbig, visiting
from the University of California, was interesting for, among
other things, a missed opportunity. The subject was the eclipsing
binary star YY Geminorum. It is one of three stars, each of them
binary, which make up the system of the bright star Castor. Castor is thus, in fact, a sextuple system. They reported that "the
striking variation in the emission line intensities discovered by
[A. H.] Joy and [R. F.] Sanford in 1926 was not noticeable in
1948–49."[103] Later investigations of YY Geminorum, some of
them carried out at McDonald, have proved it to belong to a class
of stars which have transitory spots on them, rather like sun-
spots, but on a much larger scale, and much longer lived. These
stars also show very short-lived but violent outbursts of radiation,
similar to solar flares, but on a much larger scale. For once,
Struve missed an opportunity to be the originator of a whole new
line of research, since the star seems to have been quiescent during his observations.

Struve and Herbig also observed a number of very short period
pulsating variable stars such as CY Aquilae and AI Velorum, with
periods in the range of a couple of hours. This set a precedent for
the investigation of a type of star which aroused intense interest
among their successors.

The list goes on and on: Struve had an uncanny knack for picking for his own work or recommending to his staff topics of far-
reaching import. The list could hardly have been better chosen if
he had been able to foresee the future. Whether he could foresee
it or not, he was certainly making it.

Otto Struve died in 1963. It is fitting that, in the early summer
of 1966, a further ceremony was held at Mount Locke, accompanied by a symposium at Marfa. The 82-inch telescope was re-
named the Otto Struve Memorial Reflector. Henceforward, the
observatory would be William J. McDonald's memorial, but the
telescope belonged to the memory of Otto Struve.

CHAPTER 1 2

Interregnum

The administrative snarl caused by the abrupt resignation of Kuiper and the departure of Struve came to an end when Bengt Strömgren agreed to return to Chicago as chairman of the Astronomy Department and director of the Yerkes/McDonald Observatories, effective 1 January 1951. Shortly before Strömgren took up his new duties, the vacancy left by the resignation of Paul Jose was filled by the appointment of Marlyn Krebs as superintendent. For over a year, the day-to-day administration of the facility at Mount Locke had apparently fallen into the hands of secretary Dorothy Hinds. Krebs' arrival at Mount Locke filled a local vacuum, while that of Strömgren at Chicago resolved an institutional diffusion of authority by rejoining the directorship to the chairmanship.

Neither man was a stranger to the observatory. Strömgren, of course, had already served there as an astronomer under Struve. Krebs was the son of the long-time superintendent at Williams Bay and had grown up under the eyes of many of the astronomers who would come to Mount Locke; W. W. Morgan had been his Boy Scout leader. The younger Krebs' background of instrument-making, engineering, and watchmaking would stand him in good stead for the next decade and a half on the mountain. What Krebs was *not* was an astronomer. The post of resident astronomer was put on the shelf, to be resurrected only briefly in 1959 and then abandoned permanently. This was compensated by the expectation that the assistant associate director would spend a large fraction of his time at Mount Locke. When neither he nor the director was there, another of the astronomers visiting from Williams Bay was designated to "act for the director." For the next decade, astronomical authority on the mountain would be a sort of round-robin affair, with virtually every senior astronomer taking a turn.

One of the first priorities of the new administration was to improve the comfort of living on Mount Locke. All of the houses were renovated; the bare wood and papered walls were covered with insulating plasterboard, attics received further insulation, windows were replaced with thermopane glass, and the primitive butane heating units were supplanted by thermostatically controlled wall heaters. Winters were going to be a lot less painful from now on. The uneven dirt paths were provided with concrete walkways to reduce the hazards of walking at night, hazards that had caused numerous minor injuries.

Like Jose before him, Marlyn Krebs had school-age children, and he soon became active in Fort Davis social affairs. Despite his being a newcomer and a Yankee, he was invited to run for the Fort Davis School Board and was promptly elected to its membership. He was instrumental in expanding the scope of vocational training in the high school, and—in the wake of the 1954 Supreme Court decision—played an active part in ending the separate but unequal school facilities for whites and Hispanics. (Since the departure of the black cavalry unit that manned the active military fort in the nineteenth century, there have been virtually no blacks in Jeff Davis County.) Krebs remained on the School Board for twelve years.

A Warner Brothers Production

There were some important instrumental innovations during this period. Besides producing an infrared photoelectric photometer, Hiltner finished a new coudé spectrograph, which was installed with the aid of the newly arrived Marlyn Krebs in April 1951. This instrument used a choice of large diffraction gratings to disperse the light and a series of Schmidt-type reflecting cameras mounted on an axis allowing the chosen one to be turned into position, to focus the spectra. Hiltner and Arthur Code worked on a photoelectric spectrum scanner for the Cassegrain focus, to record spectral line intensities directly, without the intervention of a photographic plate. In addition further experiments on image intensification were conducted. Meinel completed the installation of 8-inch and 10-inch coudé spectrograph cameras in 1954.

The original chromium and aluminum coating of the 82-inch mirror had stood up extremely well. Toward the end of the war, when some of the smaller mirrors were recoated in the vacuum chamber built under Hiltner's direction, the main mirror was still in such good condition that it was simply cleaned and reinstalled. Even by the fall of 1950, its reflectivity was only diminished by 10 percent, but it was decided to recoat it then.

The mechanical support system for the 82-inch mirror was still posing problems. By 1953, a new support cell had been designed by the firm of Joseph Nunn and Associates. Most of the fabrication work on the new cell was done by John Vosatka at Yerkes. This massive piece of steel arrived by rail at Kent, forty miles north of Mount Locke. Marlyn Krebs, Tommy Hartnett, and Joe Rodríguez were able to manhandle it onto the observatory truck, but the problem of getting it into the dome and onto the telescope defeated them.

Just then, Warner Brothers were making the movie *Giant* on the Worth Evans ranch south of Marfa. They had a very large crane for the construction of their sets, including the façade of a large house that was a local landmark for many years. Krebs visited the set and Worth Evans introduced him to director George Stevens and rigging foreman Myron Shindler. Mrs. Krebs invited them to the house for dinner, after which they were given the red carpet tour of the observatory. At the telescope, Krebs mentioned his problem, and it was agreed that the Warner Brothers crew and crane would generously help a different kind of star by lifting the cell into the dome.

There was a small problem with the large crane. It was not licensed to travel on the highway, nor to pass through the towns. Nor could it be. It was overweight and overlength, posing a potential hazard to the numerous cattle guards along the route. Very early one morning, the rigging crew rumbled their behemoth northward through Marfa and Fort Davis, hoping not to be noticed. Meeting no challenge, they inched their way up the mountain and accomplished their mission.

They had been noticed, though. A telephone call from a friend in town revealed that the State Highway Patrol was lying in wait for them at the northern edge of Fort Davis. No problem. Instead of taking the 16-mile direct route, the crane was driven around the 50-mile Davis Mountain Scenic Drive, circumnavigating Fort Davis. When Krebs drove back through town in an observatory truck about noon, the police were still there, emptyhanded. They stopped Krebs to ask about the crane at the observatory. "There's no crane up there," they were told. A call to the Worth Evans ranch brought the equally innocent declaration that "Oh, no, the crane has been here all day! Come see for yourself." Instead, they followed Krebs up the mountain, just to make certain.

A few days later, *Giant* star James Dean drove up the mountain to look around. It was after hours on a Sunday, and everything was locked. The Krebs were getting ready to drive to town when Dean found Marlyn Krebs and insisted on seeing the telescope. Krebs refused and said that he would have to come back during the usual hours. As they were driving down the mountain, the superintendent told his wife about the encounter. "You didn't turn James Dean away?" she asked incredulously. "Who is he?" was the reply. Dean later got his tour, by invitation.

The Nunn firm also designed the mechanical parts of a new 36-inch reflector, which was built by the Boller and Chivens Company. It was the first exercise in telescope construction by this now-famous manufacturer. The 36-inch telescope is an unusual instrument, superficially resembling a conventional Cassegrain configuration, but technically known as a Dall-Kirkham, in which the primary is ellipsoidal in shape and the secondary a convex sphere. The mirrors were made in the Yerkes optical shop. The building to house it was constructed to the southeast of the 82-inch dome and lower down the mountain, near Benedict Bench. Mount Locke employees Tommy Hartnett and Eddie Webster laid the walls from stone that they quarried from the Eppenauer ranch, while Webster and Krebs welded the dome. An automobile garage lift was bought as the foundation for the elevator floor, and the crew jury-rigged their own makeshift crane to get the telescope mounting into the completed building. "We were lucky that no one got hurt on the job," Krebs remarked recently.[104] In fact, he himself was the only casualty; he suffered deafness in one ear, from an accident during the blasting away of a piece of the mountainside to make a hole in which to put the building.

The 36-inch instrument went into operation in February 1957. It has been used primarily for photoelectric photometry, for which it has proven to be admirably suited. Boller and Chivens made a duplicate, including the optics, for the Leiden Observatory Southern Station at Hartbeespoortdam, South Africa, and another for the University of Wisconsin.

Starlight, Star Bright, What Color Are You Tonight?

Prior to the acquisition of the 36-inch, Hiltner had built a 13-inch reflecting telescope for photoelectric photometry. The chief beneficiary of this instrument was Harold L. Johnson, an assistant professor at Chicago, the first time around, from 1950 to 1953. In 1957 he was appointed to a tenured associate professorship, and later he transferred to Austin.

Johnson was to have a profound influence on the progress of observational and theoretical astrophysics, despite an early death. His work was mainly confined to photometry, initially the development of what is called the UBV system, based on observations made by himself and W. W. Morgan, and referred to the spectral classifications established by Morgan. This system consists of measurement of the brightness of stars (or other celestial objects) through selected filters of well-defined color transmission. The three passbands of the UBV system, as defined by Johnson's work, are in the ultraviolet (U), blue (B), and visual (V) or red-yellow regions of the electromagnetic spectrum. The differences in magnitude measured for the three colors (i.e., U-B, U-V, B-V) represent a star's *color indices,* which define its observed color.

Star colors are influenced primarily by the surface temperature of the observed object, but they are also seriously affected by the characteristics of the interstellar material through which the light has passed on its journey from the stellar source to the Earth. The intrinsic color of a star is also related to its luminosity, which in turn depends on whether the star is "normal," a supergiant, a giant, a dwarf, or one of the other denizens of the celestial zoo. More subtly, it can even tell something about the non-hydrogen content of the star's atmosphere. Color systems such as the UBV are powerful diagnostic tools for stellar classification.

Johnson's UBV system was first to be adopted worldwide, just as photoelectric photometry became a standard astronomical technique. In later work, Johnson extended the UBV photometric system to include a number of other passbands in the infrared region, requiring newer observational techniques. Others have been added by his successors. Consequently, the majority of catalogues of stellar colors have been cast in terms of Johnson's work. Although Johnson's work neither began nor ended nor was continuous at McDonald Observatory, the continuity of the work makes it appropriate to treat it as a whole. Its impact throughout astronomy calls for an extended summary.

Johnson and Morgan calibrated the UBV magnitude system in terms of stellar spectral characteristics and luminosity class. This work was later extended by Daniel L. Harris and Adrian Blaauw and published by them in the stellar volume of the great nine-volume compendium of modern astronomy, *Stars and Stellar Systems,* which Kuiper conceived in 1955 and for which he served as general editor. This standard work took many years to produce, the final volume on galaxies only appearing in 1975, two years after Kuiper's death.

One of the important aspects of this work was unraveling the effects of *interstellar reddening* and *extinction.* The space between the stars is not an empty vacuum—not quite. There are clumps of both gas and dust interspersed all over. When starlight passes through dust, it becomes redder. This is a common enough effect; it is exactly the same mechanism that makes red sunrises and sunsets. Dust grains tend to scatter the blue wavelengths more strongly than the red, thus dimming the light to some extent in all colors, but especially in the blue part of the spectrum. Determining the true properties of stars requires that the loss of light in various colors be corrected in the observations. The point of the UBV calibration was to permit such corrections to be determined directly from the photoelectric measures.

Perhaps the most important application of the new technique was to the study of clusters of stars. The members of a star cluster are all stars formed at about the same time and from material having the same composition. The work of Johnson on such nearby clusters as the Pleiades, the Hyades, and Praesepe demonstrated observationally the effects of stellar evolution. These are the changes that take place as stars consume their material by nuclear transformation, the process that causes them to shine. Theoretical predictions of the initial effects had already been made by Chandrasekhar and his colleagues. Johnson's observational work formed the basis for the further development of theories of stellar evolution and the assignment of ages to different clusters.

Johnson left Yerkes/McDonald to join the staff of the Lowell Observatory in 1953, but he maintained a close association with his former colleagues and continued his use of the McDonald telescope.

Bengt Strömgren's name is associated with the concept of the Strömgren sphere, the volume of the cloud of interstellar hydrogen capable of being ionized by the shortwave radiation from a hot star. He, too, was experimenting with a photometric system, of a different sort from Johnson's, but also based on the use of a series of narrow band filters. Combinations of measurements of starlight passing through them are used to classify the star according to its luminosity and to determine the proportions of different elements, compared with the hydrogen and helium content.

Hydrogen is the simplest of elements, each atom containing only one proton and one electron. It and helium are believed to have been the only elements present in the beginning. Additional helium is produced from hydrogen by the simplest stellar thermonuclear process. All heavier elements, loosely called "metals" in this context, are produced by more complex processes at a more advanced stage of stellar evolution. The proportion of metals to hydrogen and helium is always very small, about 2 percent for our Sun, for example. If this proportion is lower than average, the star can be diagnosed as of great age, formed before the interstellar medium became enriched with heavier elements produced by other stars in the course of their evolutionary changes. Higher proportions of these elements indicate more recent condensation, a young star. Strömgren's photometric system thus addressed the question of individual stellar ages.

Narrow-band photometric studies by Hiltner and Strömgren on the distribution of hydrogen in our own Milky Way galaxy inspired the latter to conclude that, in general, there is a density of about 0.2 atoms and ions per cubic centimeter, but that there are isolated clouds where the density might be as high as 50 per cubic centimeter. These results would later be confirmed and extended to far greater detail by radio astronomers. Such densities are, of course, far lower than the best terrestrial vacuum attainable, but Strömgren's discovery showed that the interstellar hydrogen contributes an important part of the total mass of the galaxy, because of the gigantic volumes occupied by these tenuous gases.

Margaret and Geoffrey Burbidge, who were then interested in high-dispersion (fine resolution) spectroscopy of stars, spent the 1951–1952 academic year at Yerkes/McDonald and made heavy use of the 82-inch telescope. They returned to regular staff positions at the University of Chicago in 1957. By then, they had turned their attentions to studies of extragalactic nebulae, particularly the galactic rotation work that had been initiated earlier by Thornton Page.

On the administrative side, Aden Meinel became associate director of the observatories in June 1953, to be succeeded by

Adrian Blaauw as associate director of McDonald in 1956. A year later, Blaauw left that position to become director at Groningen, in his native Netherlands. For the moment, the position at McDonald remained vacant.

The departures of Meinel and Blaauw reflected a larger reality. A center of excellence such as Yerkes/McDonald/Chicago was during the postwar decade attracted astronomers from far and wide. Once at the peak, though, there was only one way to go. Concentrations of talent attract talent from elsewhere, but they also attract offers of employment to the most able individuals. Other institutions were anxious to employ staff of the highest caliber. Astronomy was experiencing a time of expansion, with the development of radio astronomy, the establishment of the Kitt Peak National Observatory, and the inception of massive astronomical activity in Arizona, Hawaii, and Chile. Eventually, many of the great McDonald observationalists of this era would go to direct other and newer enterprises.

After seven years, even the mild and debonair Strömgren wearied of the effort required to oversee his lively and farflung organization. He resigned his three simultaneous administrative posts, effective 1 September 1957. Gerard Kuiper was invited to try a second term, this time with real directorial powers. Until now, there had been relatively little participation by personnel from the University of Texas, but Kuiper was determined to change that. The time was approaching when that institution would have to make critical decisions concerning its future astronomy program and participation in the work of McDonald Observatory. Whether he intended it or not, Kuiper's drive for a viable astronomy department in Austin sealed the fate of the joint Yerkes/McDonald operation. Its days were numbered.

CHAPTER 13

Texas Independence

Astronomy Takes Root in Austin

The original Chicago-Texas compact that had been so carefully crafted by Struve and Benedict was due to expire in 1962. Both parties had foreseen that any new agreement would be markedly different in character, and this was reflected in events at both institutions.

The approach of the expiry had an effect probably unanticipated by Struve. Neither university was willing to provide adequate funds to keep McDonald really competitive. In Austin, the opinion was that Chicago astronomers were profiting from the facility, so Chicago should pay the expenses. Conversely, during the decade of the 1950's, the work done at the McDonald telescope formed a steadily less prominent part of Yerkes/Chicago activities. At least in the later stages, the prospect of investing large funds for new instrumentation and maintenance in a facility that they did not own and might not control much longer worried the Chicago administrators. They began to look elsewhere and to be prudent in their expenditures.

It was clear that the University of Chicago had been reducing its financial commitment to McDonald over a period of several years, with the apparent intent of cutting its losses in the event of a total abandonment of the cooperation. Whether it was a cause or an effect of this, the University of Texas was at the same time building its own astronomy department toward eventual self-sufficiency.

In a sense, that department was partly a creation of Otto Struve. It is true that Dean Benedict had inspired the transformation of the Mathematics Department to Mathematics and Astronomy in 1925, a year before McDonald's death, but this meant very little in practical terms. Efforts by Benedict and Professor E. J. Prouse to start an undergraduate astronomy program were abortive. Prouse was a member of the mathematics faculty, but had studied astrophysics under C. D. Shane at the University of California. For many years, he taught the only astronomy courses (listed as mathematics, before 1925) in the university.

Struve's contribution was to recommend one of his recent students to UT's President Theophilus S. Painter. Frank N. Edmonds, Jr., had taken a position at the University of Missouri after being awarded a doctorate by the University of Chicago in 1950. Two years later, he was appointed to the Texas faculty of Mathematics and Astronomy in Austin. By that time, Struve had been replaced by Bengt Strömgren as director of the McDonald and Yerkes observatories, but he too encouraged and supported Edmonds' efforts as the beachhead of an astronomical presence in the Texas capital. The new Austinite participated fully in the collaboration, undertaking high-dispersion spectroscopy of the star Procyon with the 82-inch reflector. In Austin, he continued his mathematical studies of stellar atmospheres.

Gerard Kuiper became director of Yerkes and McDonald observatories for the second time on 1 September 1957, succeeding Strömgren. He soon proposed the establishment of a joint Department of Astronomy, from which graduate students could receive advanced degrees issued jointly by Chicago and Texas. As a result of this novel idea, the University of Texas divided the Department of Mathematics and Astronomy created in 1925 by H. Y. Benedict into separate Departments of Astronomy and Mathematics, effective in the fall of 1958. Edmonds became associate professor of astronomy, the department's first faculty member. Prouse remained in Mathematics, but continued to teach astronomy in collaboration with the new department, and he even spent a brief while working at Yerkes. Pending a formal protocol with Chicago, there was no astronomy chairman for the moment.

It was a period of turmoil in Williams Bay, and the remote facility on Mount Locke was not always uppermost in the minds of Yerkes astronomers. They had been accustomed to a very satisfying degree of autonomy, with very little interference from the Chicago campus. That was coming to an end. The university administration was beginning to curtail the freewheeling astronomers in irksome ways. Add this to the fact that many of them were strong and eccentric personalities—common enough in astronomers—and the result was apparently a lot of friction, short tempers, and short directorial tenures. Especially Kuiper's.

Kuiper resigned abruptly as director of the Yerkes and McDonald observatories for the second time, early in 1959. He was, according to Chandrasekhar, fed up with subordinates who "were rude and disagreeable towards him."[101] Like Struve before him, he retained the department chairmanship until the following year. The irascible Hollander was succeeded in both posts by William W. Morgan, the last Chicagoan to pass through the revolving door of the joint directorship in the post-Struve years.[105] The winds of change became even more evident when Austin's Frank Edmonds was appointed associate director of the McDonald Observatory, under Morgan, effective 1 March 1960, disappointing Johnson and Hiltner.

There had been no resident astronomer on the mountain for eight years. Morgan brought the post back in 1959, appointing Jurgen Stock. The fragmentation of authority between Stock and Marlyn Krebs didn't work at all, however, and Stock was soon dispatched on a site-testing mission to Chile. While he was gone, the position was abandoned for good.

Chandrasekhar cautions that the conflicts within the Chicago astronomy group have often been exaggerated, especially in Austin. It is clear that close-knit camaraderie did exist much of the time. As in associations of faculty and graduate students everywhere, there were boisterous parties and arcane silliness, such as the spontaneous party songs "W. W. W. W. Morgan" (which has perhaps fortunately been lost) and "The Billy Bidelman Song":

(tune: "Battle Hymn of the Republic")
Struve, Kuiper, Hiltner, Morgan, Chandrasekhar too,
Struve, Kuiper, Hiltner, Morgan, Chandrasekhar too,
Struve, Kuiper, Hiltner, Morgan, Chandrasekhar too,
And Billy Bidelman.

No one seems to recall why William P. Bidelman deserved such august company, but it must have seemed a good idea at the time.

It is an interesting commentary on the contemporary state of astronomy in the United States that McDonald did not have an American-born or American-educated director until nearly three decades after its founding. Of Morgan's predecessors, only Bengt Strömgren, son of the well-known Danish astronomer Elis Strömgren, was not a U.S. citizen. Nonetheless, Struve was born Russian, Kuiper Dutch, while Cambridge-educated Chandrasekhar was Indian. Since Morgan was born in Tennessee, there is no record of regental objections about Yankees.

Following in Struve's footsteps, Kuiper retained the Chicago departmental chairmanship and thus the de facto cochairmanship of the joint department in Austin. At first the corresponding cochairmanship in Austin was vacant. Harold Johnson was appointed professor of astronomy at Texas in 1959, and the formal operation of the joint department began with the adoption of bylaws on 28 June 1960, with Johnson and Kuiper as cochairmen. Shortly afterward, Kuiper left Chicago to found the Lunar and Planetary Laboratory at the University of Arizona. He remained there until his death on 24 December 1973.

Gerard de Vaucouleurs was appointed associate professor on 1 October 1960, bringing strength to extragalactic programs at Texas. De Vaucouleurs, a Frenchman by birth, had worked at Mount Stromlo Observatory in Australia, where he had studied the Magellanic Clouds and bright southern galaxies. He was also for a time at the Lowell Observatory and at Harvard, concerned with the topography and physical conditions on the planet Mars. It was during this time that he formulated the concept of the Local Supergalaxy, that there exists a cluster of many galaxies, including our own Milky Way, with coordinated motions. These systematic motions would affect determinations of the parameter (the Hubble constant) describing the large-scale expansion of the Universe, thereby causing the size of the Universe to be overestimated. De Vaucouleurs would be awarded the prestigious Herschel Medal of the British Royal Astronomical Society for this discovery and for his subsequent detailed studies of the Supergalaxy, two decades after his arrival in Austin.

The joint department arrangement didn't work and was soon abandoned. Harold Johnson became chairman of the Astronomy Department at Austin on 1 September 1961. He had a more mercurial temperament even than most astronomers, waxing alternately enthusiastic and pessimistic on a time scale of a few days. He also thought that the chairmanship carried on implicit promise of the McDonald directorship under the new agreement being forged between the two universities. To his surprise, no one else saw it that way. Due to differences with the university administration, he resigned the next February, to join Kuiper at the Lunar and Planetary Laboratory at Tucson. Edmonds became acting chairman, yielding that post to Gerard de Vaucouleurs when he spent the 1962–1963 academic year abroad on a Guggenheim Fellowship.

Thus, with fits and starts, did astronomy begin to get a firm foothold in Austin, building toward the day when the McDonald Observatory could look eastward for support and control, rather than to the north. During the period 1958–1963, a real astronomical presence began to take form in Austin, with help from the university administration, the National Science Foundation, and the Office of Naval Research.

Not everyone was easily convinced. When Otto Struve's dream of a national consortium observatory was realized with the foundation of the Association of Universities for Research in Astronomy (AURA), a consortium invented to manage the Kitt Peak National Observatory, Chicago became a member immediately. Texas was spurned, however, despite its ownership of McDonald Observatory. This slight was reversed only after strong representations by Austin's Dean of Graduate Studies, W. Gordon Whaley.

Mars and the Galaxies: Research, 1957–1963

Despite the administrative problems during this period, a considerable amount of frontier research was still being done on Mount Locke. Kuiper's survey of asteroids with the Cook telescope was completed and published. The *Rectified Lunar Atlas,* mentioned earlier, was also published. The Mars opposition of 1958 brought the planet closer to Earth than usual. It was an uncommonly good opportunity to study both the planetary surface and the satellites. Kuiper took full advantage of the opportunity and not only did a lot of science, but also got some publicity for what the press called a search for extraterrestrial life, such as lichens on Mars. One source alleges that Kuiper invited a selected group of reporters to be with him in the dome on a certain night, when he proposed to make a momentous discovery.

The aging but indefatigable van Biesbroeck continued his observations of comets and double stars, as well as undertaking measures of earthlight in the atmosphere as part of an International Geophysical Year program. Part of the time, he worked with his visiting southern hemisphere counterpart in double star astronomy, Willem van den Bos, emeritus director of the Union Observatory of South Africa in Johannesburg.

Robert Kraft undertook one of his classical investigations of Cepheids with the object of settling the vexed question of the color excesses of these stars. The fact that these variable stars all occur in the galactic plane, where interstellar extinction and reddening are always appreciable and sometimes extreme, has often dashed hopes of using them as precise distance indicators even though their absolute luminosities are very closely related to their times of periodic variation. Kraft's numerous studies over many years eventually led to a resolution of this problem.

Harold L. Johnson rejoined the Chicago faculty in 1958, transferring to the Austin faculty the following year. Shortly after arriving in Austin, he introduced the computer reduction of photoelectric observations. By 1960, he was observing classical Cepheids in nine different band passes. In addition to the original UBV wavelengths, he had now added infrared bands R, I, J, K, L, and M at 680, 820, 1300, 2200, 3600, and 4700 nanometers respectively, passing well beyond the range of visible light. Johnson was more concerned with making observations than using them, so the gestation period from telescope to conclusion was longer than in many cases. Two decades later, his observations permitted a definitive determination of the Cepheid distances (see Epilogue).

Another photometric observer provided one of the livelier moments on the mountain during this period. John S. Neff was a postdoctoral research fellow at Yerkes in the early 1960's and used a large fraction of the observing time at the 36-inch reflector. In those days, the 36-inch dome was where the road stopped, and there was not yet a retaining wall. Where the road stops, the mountainside plunges precipitously for nearly a thousand feet. One summer night in 1963, Neff drove off the edge. Luckily for him, he fell out of the car on its first bounce. It continued bouncing far down the mountainside. For a while after, Neff called that slope "my shortcut to Fort Davis," while Krebs still calls it "Neff's Landing Strip." Ignoring the fact that he wouldn't have driven over the edge in daytime, Neff remarks, "It's a pity that it was dark and I couldn't see the car. It must have been spectacular."

This period is most notable for the development of extragalactic research at McDonald. When Margaret and Geoffrey Burbidge came to Chicago permanently from Britain in 1957, they were chiefly interested in stellar topics. Now, however, they followed in the footsteps of Thornton Page by using the B spectrograph at the prime focus of the 82-inch for a long series of investigations of the rotation of galaxies. This work relied on the fact that the B spectrograph had a very long slit, which could cover the whole extent of a medium-sized galaxy. This was not possible with most spectrographs. These data were used in analysis of mass distribution of galaxies. They also obtained large numbers of direct photographs of galaxies. This important program, which also involved several junior collaborators, continued until 1962, when the Burbidges transferred to the University of California at La Jolla.

With his wide interests and energetic approach, Gerard de Vaucouleurs, assisted as always by his wife, Antoinette, quickly began to have a profound influence on the Texas research program. He continued a program, begun at Harvard in 1959, of photometry of bright southern galaxies, using photographic material obtained during his earlier stay at Mount Stromlo. This was amplified by standardized photographs of galaxies obtained at McDonald and UBV photometric observations of many systems. He also analyzed the structure of the Virgo cluster of galaxies, determined new redshifts and, together with his wife, embarked on the compilation of the *Reference Catalogue of Bright Galaxies*. With the aid of several collaborators, studies were conducted of the absorption within the Milky Way, and a major project was launched for the determination of reliable magnitudes and colors for about a thousand galaxies north of declination −35 degrees. As this coincided with the Burbidges' work on galaxy rotation and studies of the mass-to-light ratio of various types of structure within galaxies, and with Morgan's work on a novel system of galaxy classification, the Yerkes-McDonald-Chicago-Austin combination became one of the leading centers in the world for the study of galaxies.

At the same time, Gerard de Vaucouleurs was engaged in detailed mapping of the Martian surface from visual observations. There are not many astronomers alive today who can match his competence for standing at the eyepiece of a telescope with only his bare eyeball and recording the details of what he sees. He doesn't do it anymore, but the accuracy of what he did with Mars in the 1950's and 1960's was spectacularly validated by the cameras of the spacecraft that visited that planet in the 1960's and 1970's.

Coincident with the strengthening of the Austin department, the armory of ancillary equipment at McDonald was being improved. Robert G. Tull, an expert on spectrum scanners and spectrographs, had joined the faculty. Together with Johnnie E. Floyd, who would be McDonald's chief design engineer for many years, he designed a spectrum scanner for the Cassegrain focus of the 82-inch reflector.

The Second Texas-Chicago Agreement

In 1962, the University of Texas and the University of Chicago celebrated their thirty years of collaboration by signing a new agreement governing the operation of the McDonald Observatory. It was quite different from that hammered out between Otto Struve, working through Robert Hutchins, and Harry Yandell Benedict.

The history of the second agreement is one of an institution still trying to decide where to go and of another fearful of the eventual decision. The Texas deliberations and Chicago rumors began a full five years before the expiry of the 1932 compact. In Austin, sentiment among the administrators flopped back and forth between several extremes. Should they renew with Chicago, but with a Texas director? Why not initiate an open competition for a collaborative partner, in the style of the original Chicago agreement, but with what amounted to competitive bidding? Harvard in particular was mentioned in the latter context. Even total abandonment of astronomy was considered possible; a rumor reached Morgan that Texas Provost Norman Hackerman had asked one astronomer, "You people can find jobs elsewhere, can't you?"

In Chicago, the rumors got to the point that Subrahmanyan Chandrasekhar wrote to Dean Whaley about the problem. Chandrasekhar had just been elected to the graduate faculty of the University of Texas as part of the arrangements for the joint department. He explained that, despite having no personal stake in telescopes, he was immensely concerned about the future arrangements. In effect, he said that Chicagoans did not know if the rumors were true, but that if Texas thought it in their best interests to abandon Chicago, Texas at least owed Chicago the right to be informed as quickly as possible. It was an admirable letter.

It was an effective letter, too. Whaley quickly assured everyone that Texas had no intention of abandoning either Chicago or astronomy. The rumors that had flown about among the Chicago and Austin faculty were squelched. The hopes of several individuals who had been convinced that each should be the new director were disappointed when Texas proposed a search committee. From then on, progress was rapid. The new agreement was negotiated. Some of the Yerkes people were displeased with the terms, but this was just an expression of the more general dissatisfaction that had led to the rapid alternations of the directorship over the years.

The new agreement was written in March 1962, to become effective the following November. As major changes from the preceding compact, it provided (a) that the agreement would have a ten-year lifetime; (b) that the director would be appointed by the University of Texas, upon the advice of an external committee of astronomers and after consultation with the Yerkes director; (c) that Texas and Chicago astronomers would receive equal consideration for observing time, with each institution guaranteed at least 25 percent of the time available and not more than 25 percent to be given to other institutions; (d) that eventually Harvard/Smithsonian could be made a partner to the agreement, with the consent of the directors and administrative officers of both Texas and Chicago; (e) that observing time on instruments not yet built was not covered by the agreement, but subject to later negotiation; and (f) that since the observatory was now an integral part of the University of Texas, its funding was entirely a responsibility of that institution.

The new compact could not be put into full effect immediately, because the most important change was the selection of a Texas director. That process took more time than anticipated, so it was necessary to make an interim arrangment. An agreement dated 10 August 1962 provided that W. W. Morgan would remain director of McDonald until 31 August 1963 or until the appointment of a new director by the University of Texas, whichever came first.

Texas Gets Its Own Director

The new agreement provided that the University of Texas would appoint the director, independently of any administrative arrangements at Yerkes and Chicago. Henceforward, the director would report to the president of the University of Texas and be accountable in no way to the northerners.

That is not to say that Chicago opinion was to be neglected in choosing the first Texas Director. Texas Provost Norman Hackerman appointed a search committee to advise him on potential candidates. Its members were former Yerkes/McDonald director Bengt Strömgren, ex-Chicagoan Horace D. Babcock, Lick Observatory's Albert E. Whitford, and Harvard's Donald Menzel. While not a formal member of the committee, W. W. Morgan was invited to sit in on the deliberations, since Texas would consult with him for recommendations on the implementation of the report.

The committee convened for two days of meetings, the first at Indian Lodge in the Davis Mountains, the following day in Austin. They hammered out a list of seven potential candidates, one of whom was subsequently dropped before any action was taken. None of the Yerkes or Austin astronomers was considered. The list was headed by J. B. Oke, of the Mount Wilson and Palomar observatories, but even some members of the committee were convinced that Oke would not accept. This turned out to be true. "The man from Caltech turned us down flat," Dean Whaley reported. Although he was not the next one on the list, Morgan suggested to Whaley that Yale Observatory's Harlan J. Smith should be the next candidate. Morgan has since characterized this act as "one of the smartest decisions I ever made."[23]

Early in 1963, Smith was invited to apply for the McDonald directorship. His interview visit to Austin coincided with the dedication of a new millimeter wavelength telescope at the university's Balcones Research Center, beyond the northern outskirts of town. During the course of the visit, Gordon Whaley made a verbal offer of the McDonald directorship, which included the chairmanship of the Austin Astronomy Department. There was a brief encounter with Hackerman, who in his customary gruff way barked, "Remember, we just don't want to pour money down any rathole."[106]

After a few days' reflection, Smith accepted, conditional on five points. The Austin faculty should be allowed to grow sufficiently to provide both depth and breadth, as well as to be able to take advantage of the McDonald facilities. The physical plant on Mount Locke should receive adequate funding for maintenance and upgrading, so neglected by both universities in the recent past. Smith wanted permission to "promote a substantially larger and more modern telescope for McDonald." Radio astronomy should have an active role, cross-fertilized with the optical work. Finally, there should be eventual support, when it became appropriate, for expanding into space astronomy.

The letter containing these points received no written reply. Characteristically, Hackerman telephoned Smith at a vacation retreat on Martha's Vineyard at 7 A.M. (6 A.M. Austin time) on a late June morning to solidify the deal. Texas had its man.

Morgan resigned the McDonald directorship effective 1 September 1963. The development of the McDonald Observatory as an autonomous facility of the University of Texas, though still shared with the University of Chicago, begins on that date. Just as Otto Struve's conditions, embedded in the original interuniversity agreement, brought McDonald to astronomical preeminence, Smith's conditions would shape the future in ways that reversed the slow downhill slide of the 1950's and justified independence.

CHAPTER 14

Launching the Modern Era

The conditions under which Harlan Smith accepted the position of director implied an ambitious development program, both at Mount Locke and in Austin. He now put it into effect. Over the next few years, there would be less research, but much progress in the research capabilities of the McDonald Observatory.

Texereau of Texas

One of the first tasks to be undertaken was the improvement of the performance of the 82-inch telescope. J. S. Plaskett's 1939 account of his Hartmann tests, based on reflections back to the center of curvature from different places on the mirror surface, had claimed a figure of extraordinary accuracy, with a maximum error of only 5 percent of a wavelength of visible light. Once the telescope was on the mountain, no astronomer had been able to verify Plaskett's claim. It was clear that the support cell was badly designed, and several attempts were made over the years to improve it. Struve had been critical of the care put into the fabrication of the secondary mirrors, but even the prime focus images— which use only the primary mirror—were less than perfect.

The secondary mirrors, one for Cassegrain and one for coudé, were much smaller than the 82-inch primary, so they would be easier to work with. Even if the problem were in the primary, a first-class optician could shape the secondaries to compensate for the faults of the big mirror, at much less cost than reworking the primary.

At the suggestion of Gerard de Vaucouleurs, Smith called in Jean Texereau, a highly-skilled astronomical optician from France, with a scarcely less impressive expertise in the appreciation of fine wines. Yerkes colleagues joked that, if McDonald had Texereau, they should be able to find an expert called Chicagereau.

A quarter of a century later, the elfin Frenchman's eyes sparkled as he told the story of his adventurous drive from Austin to West Texas. During the night, his car broke down somewhere west of Sheffield. From that small village to the next town of any size, Fort Stockton, was an empty 75 miles. A Sheffield garagist telephoned to the observatory, still a hundred miles beyond Fort Stockton. Eventually, an observatory car arrived in rescue. Texereau's Mercedes was not salvageable that night, so they drove into town, to register the abandonment with the Pecos County sheriff's office. Texereau and his rescuer walked into the building at about 2:00 A.M., to be confronted with "a scene out of a movie. A deputy sheriff sat at the desk, cowboy hat on his head, feet on the desk, and a revolver beside them."[76] John Wayne's surrogate showed the courtesy that is endemic to West Texas and agreed to take care of the car in the morning. The astronomers continued on to Mount Locke.

On the mountain, Texereau met Krebs, whom he characterized as "the man who fixed everything," and the work crew of Tommy Hartnett, Lalo Granado, and newcomer George Grubb. He would work with them for fourteen weeks.

Texereau explained his problem in the following way. "Mirrors of up to 80-inch diameter have usually been finished in the optical shop only. They are generally credited with wonderful accuracy, but . . . this ignores flexure and thermal effects, which are not negligible in a telescope. Furthermore, Cassegrain combinations very often show unexpected defects, which are introduced in testing the convex secondary mirror independently. These problems can be solved simultaneously by conducting the last figuring on the secondary only, and by testing the whole combination under working conditions with a star as a point source."[107]

Texereau found serious errors in the combined system, beginning with a spherical aberration amounting to one and one-half wavelengths of light in the primary mirror. By hand working the 28-inch Cassegrain secondary in seventeen daily steps, he brought the combination to a superb condition. Short exposures on close double stars produced images near the theoretical limit, with more than half the light from a given star falling within a circle of about 0.15 arcseconds. Before he could deal with the coudé secondary mirror, he had to remove defects from the two flat mirrors (18.5 and 13.75 inches diameter, respectively) which form part of that optical train. He brought these to true flatness with an accuracy of one-thirtieth of a wavelength of light, better than one-millionth of an inch. He then found the 21.5-inch coudé secondary to be overcorrected by more than three wavelengths and badly distorted in its cell. It took twenty-one figuring steps to bring the coudé train up to the same quality as the Cassegrain.

During this long grind, while Texereau tested star images at night and worked on the mirrors by day, Harlan Smith traveled to Europe on business. In view of the aridity of West Texas, and knowing Texereau's reputation for the grape, he decided to visit a favor on the Frenchman toiling so far from his homeland. He bought what seemed to be an excellent bottle of wine to take back to him. A few days later, he arrived at the mountain. Solemnly, he sat the bottle on the table in Texereau's cottage. Texereau looked at it carefully and disappeared. A moment later, he was back, an identical bottle in hand. "Where did you get that?" Smith asked. "El Paso." Then the Frenchman pointed out that his bottle was marked "appellation controllé," whereas Smith's was not. "Leave the room," Smith commanded. While the optician was absent, Smith opened both bottles and poured equal shots into two glasses. Asking Texereau to return, he requested a judgment. One sip from each glass, and the glasses were posed, correctly, beside their parent bottles. Voilà!

They Shoot Astronomers, Don't They?

The initial period of Harlan Smith's directorate was marked by a relative paucity of scientific contributions. Only about ten papers were published per year during the years 1963–1969, compared with more than twice that many during World War II and a dozen-fold greater at present. This implies no fault to Smith, but only reflects the problems of the changeover to a fledgling Texas astronomy department and of the priorities of instrumental improvement and construction.

Even before the transfer of responsibility, the Yerkes astronomers were making less extensive use of the 82-inch telescope, partly as a consequence of the departure of some of their most assiduous observers. Many of the *McDonald Contributions* during this period represented analyses of observations made fairly far in the past, rather than presenting new research on Mount Locke.

During the transition period, the principal "Chicago" contributors were Margaret and Geoffrey Burbidge, by then removed to the University of California at La Jolla, continuing the galaxy rotation work already described in Chapters 12 and 13. From the rotational data, it was possible to deduce the mass distribution within a galaxy, and such determinations were made for numerous galaxies. Their results with the 82-inch McDonald telescope were already pointing toward one of the prime cosmological problems of the 1980's, that there seems to be far more mass in the galaxies than is indicated by their optical extent.

One casualty of this period was much regretted by everyone. Despite his efforts on the Fort Davis School Board, Marlyn Krebs had agonized for years over the problems of educating his children. They were now approaching adolescence, and he felt that he simply had to get them into more urban surroundings. With regret, he accepted a job elsewhere and tendered his resignation as superintendent in 1966; it was accepted with equal regret.

Advertisements for a new superintendent yielded no promising candidates. Eventually, Krebs was succeeded by John A. Weis, whose previous experience had been as supervising engineer in the mining industry. The habits presumably acquired in that life did not sit well with either the support staff or the astronomers. For the former, every day began with a group meeting at 8 A.M.; anyone who arrived later than 7:55 was officially late for work. For the latter, supplies were doled out with an accountability worthy of a maximum-security prison. After two years, Weis was succeeded by a space physicist, Curtis D. Laughlin, who had taken his master's degree at Iowa under James Van Allen. Laughlin remains superintendent as of 1986.

While the Yerkes-Chicago activities were tailing off, efforts were underway at Austin to vitalize the young department, which by the mid-1960's numbered ten faculty and research staff. The now-independent department was very fortunate in the quality of some of its early students. For example, Freddie Talbert did a classic study of Procyon under the eye of Frank Edmonds, Harold Ables worked on galaxy photometry with de Vaucouleurs, who also guided Rhodesian Anthony Fairall's work on properties of compact galaxies. New Zealander Graham Hill won the department's first doctorate for a study of Beta Cephei variable stars. William Kunkel used a small spectrograph fabricated by Astromechanics, a local instrument firm, to secure the first spectrum of a stellar flare taken in such a way as to provide a time-history of the growth and decay of the spectral features; flare stars would become a significant part of the McDonald program. Ronald Angione was the first to show that essentially all quasars have variable but nonperiodic brightness.

NASA support for planetary studies was evident in studies of Venus and Mars by Harlan Smith and several collaborators, including then-student Edwin Barker. The Martian atmosphere was found to be very tenuous and very dry, and it was established that the planet's surface varies enormously in altitude. Barker made the first accurate determinations of the water content of the Martian atmosphere and established its seasonal variation. Robert Roosen, using a 35-mm Nikon camera mounted on the 36-inch telescope, solved the century-old mystery of the Gegenschein, proving that this faint skyglow is a reflection from dust in the asteroid belt.

By far the most distinguished product of the University of Texas astronomy program, then as now, however, was the British-born New Zealander Beatrice Tinsley. Commuting twice a week between Austin and her home in Dallas, she took as her doctoral canvas the vast question of the evolution of galaxies, thus launching one of the important subfields of modern astronomy. While raising a pair of small children, she finished her entire Ph.D. program in the astonishingly short time of about two and a half years. She went on to an international reputation and a professorship at Yale before her tragically early death from cancer in 1981, at age 40. The University of Texas at Austin has established a visiting professorship in her honor.

Although not related directly to the McDonald Observatory per se, an important development for the department was the importing of James N. Douglas from Yale to establish a program in radio astronomy. He brought so many of his students with him that one still hears references to Texas' "Yale mafia." For the first few years, Douglas continued his already-established studies of radio emissions from Jupiter, while preparing the two-mile interferometer discussed in the Epilogue.

The Austin faculty in the mid-1960's included Andrew T. Young, a specialist in photometric instrumentation; British-born Neville Woolf, also well-regarded for his instrumentation; William H. Jefferys, a specialist in astrometry and theoretical celestial mechanics, and a former student of Smith's at Yale; and Terence Deeming, also British, a specialist in astronomical statistics.

A surprising number of astronomers are musically inclined. Deeming is a concert-quality performer on several instruments. Since the arrival of Beverley and Derek Wills, it has not been uncommon to hear violin duets drifting down from the observing floor. NASA astronomer Reinhart Beer sang to keep himself awake on long nights. Not everyone is a music-lover, though. At one point, Superintendent Weis became irritated with having his sleep disturbed, so he issued a memorandum that read in part, "Anyone caught singing loudly in the dome at night will be shot." As chance would have it, that night was cloudy, so the astronomers perversely gathered in the 82-inch dome for a group songfest. All survived, but Weis' tenure did not.

David S. Evans, who had spent the 1965–1966 academic year in Austin as a "senior visiting scientist fellow," removed from South Africa in 1968 to take on the post of associate director for research, as well as a professorship in the Austin department.

The first priority during the first few years of Texas independence was not research, however, but the refurbishing of the 82-inch telescope and the funding, construction, and equipping of a new telescope, sited slightly below and to the east of the 82-inch reflector, about halfway to Benedict Bench. To facilitate this project, engineer Charles Jenkins became associate director for administration. Jenkins' previous experience had included the same U.S. Navy test facility at China Lake, California, for which Christian Elvey had left McDonald early in World War II.

The 107-inch Reflector

In November 1962, NASA's Homer Newell contacted Dean Whaley about the possibility of contracting with the McDonald Observatory for planetary observations in support of planned spacecraft missions. Two months later, NASA branch chief for astronomy and solar physics Nancy Roman, who had done her graduate work at Yerkes/McDonald, spelled out NASA's requirements in exquisite detail. Those details were not particularly favorable to the University of Texas; they included a request that NASA have major representation on the committee that scheduled telescope time. When Smith took the reins, the NASA solicitations for major use of the observatory were redirected to him. The new director countered that the world had too few large telescopes, and that NASA could and should build one or more new ones. In the autumn of 1964, NASA signed a contract with the University of Texas for an 80-inch-class telescope to be built at Mount Locke. Somewhat later, NASA decided to build a similar facility on Mauna Kea, in Hawaii, and asked if McDonald Observatory would be interested in building it. The Texans declined.

The original agreement would have had NASA provide the entire funding for the new telescope, but this was tantamount to building a NASA "facility." NASA lawyers said that this would require an act of Congress, which is both time-consuming and politically hazardous. Consequently, Norman Hackerman, who was now the president, persuaded the university regents to accept a compromise whereby the University of Texas contracted for the new construction, with NASA providing funds for the "instrument," i.e., the telescope and coudé spectrograph, while Texas furnished the "facilities": the dome, building, piers, and peripherals. Because of its expected influence on the graduate astronomy program at Texas, an additional grant was secured from the National Science Foundation Graduate Research Facilities Office; this covered about half the cost of the dome, plus a 30-inch reflector. Rather than seeking a complete telescope from a single manufacturer, as had been done for the 82-inch instrument, it was decided to let separate contracts for the design and manufacture of the various components of the big telescope.

NASA felt an urgent need for the new telescope, in view of the impending missions to Mars. The original goal was to get observations, if possible, during the 1967 opposition of Mars, when it would be closest to Earth. The 1969 opposition was considered to be imperative. Initial inquiries indicated that a new 80-inch blank would take several years to found and anneal. This was too long. The Schott Works, in West Germany, already had a 105-inch blank. NASA agreed that the urgency of the program was such that the new telescope should use the larger mirror. When the glass blank was examined carefully by Jean Texereau, however, he found evidence of unacceptably large internal stresses. The Schott blank could not be used as it was.

By great good fortune, the Corning Glass Works had just perfected a method for the production of very large disks of pure fused silica. This material was chosen for the new telescope because of its low coefficient of thermal expansion, its annealability in only a few weeks to near freedom from internal strain, and its ease of working. The method of manufacture was extraordinary: beginning with chemically pure gases, the final silicon dioxide material was condensed from a vapor phase in a relatively small furnace to form disk-shaped boules each a few inches thick and about 60 inches in diameter. These were then sawn into hexagonal slabs and packed into a mold of the intended size of the final disk. This was then heated until the boules fused together into a single mass.

Such an elaborate process is necessitated by the fact that the fused silica cannot be melted. Under such heat, it sublimes, passing directly from a solid state into gas.

To be sure of getting a mirror at least 105 inches in useful diameter, Corning built the fusing furnace 111 inches across. Interaction of the disk consisting of the boules fused together with the mold structure required some edge trimming. By the fall of 1965, the raw disk was ready. It weighed 7,800 pounds and was 12.5 inches thick and 108 inches in diameter. It was a most remarkable sight, the material being so transparent that one could easily see through the disk from side to side and note the wavy striations where the separate boules had fused together. A hole 30 inches in diameter was cut in the center for the Cassegrain focus. Finally, the beveling of the edge reduced the finished optical diameter to 107 inches.

It is no accident that the fourth telescope on Mount Locke has a 30-inch primary mirror. Its primary came from half of the hole cut from the doughnut for the NASA reflector. The Boller and Chivens Company, makers of the 36-inch telescope mounting, modified their existing 24-inch mount to accommodate the large mirror, thereby initiating what became for them a popular 30-inch line of telescopes.

The 107-inch telescope is of the Ritchey-Chrétien pattern described in Chapter 8. Both the primary and secondary are hyperboloidal in section, the primary having a focal ratio of f/3.9 and the combination at Cassegrain focus f/8.9. Two additional secondaries give f/18 at the Cassegrain focus and f/33 at the coudé. The optical work was done in only thirty months at Davidson Optronics, West Covina, California, with opticians Charles Layton and Fred Belair working under Donald Davidson. The final figuring and acceptance tests, using the Hartmann procedure, were monitored by Robert G. Tull and Jean Texereau. The final maximum deviation of the main mirror from the desired form was no more than 7 percent of a wavelength. The mirror is capable of putting 87.3 percent of the light from a point source within a circle of 0.144 arc seconds, with no light outside one-third of an arc second. The accuracy achieved is close to the theoretical limit for a mirror of that size.

This telescope was intended primarily for high-resolution spectrographic observations at the coudé focus. Consequently, the design adopted for the mounting was a cross-axis configuration similar to that of the 82-inch instrument, with the telescope tube and counterweight on opposite sides of a polar axis supported at its two ends. The success of a Ritchey-Chrétien system requires highly accurate relative positioning of the primary and secondary mirrors. This suggested the use of a heavy, closed, telescope tube, rather than the open-girder construction used in the older design. The foci include Cassegrain and coudé configurations; the latter has such a large focal length that the southern (lower) terminus is housed in a "coudé bulge," giving the building a mildly teardrop-shaped rather than circular cross-section.

Less conventionally, there are three *Nasmyth* foci, sometimes called "folded Cassegrain." These use the Cassegrain secondary mirror, but the beam is intercepted near the bottom of the tube by the tertiary flat mirror that serves the coudé system. Instead of diverting the beam into the coudé system, the flat is rotated so as to send it out any of the three ports in the side of the telescope tube, from which the focus is accessible to an observing detector.

The final design, due largely to Charles Jones and John O'Rourke, produced a very rigid and heavy (156-ton) telescope. The contract for the construction of the mounting was awarded to the lowest bidder, the Westinghouse Electric Corporation.

The 107-inch telescope was equipped with a huge spectrograph, designed principally by Robert Tull and Johnnie Floyd, which forms an integral part of the structure of the building and the telescope. The main supports consist of a pair of parallel bridge girders each 4 feet high and 70 feet long, passing north-south through ports in the lower part of the two concrete piers. Two shorter members of similar section form a triangular frame extending eastward, with one member along the line of the light beam coming from the slit to the collimator. The parallel girders carry the two camera mirrors; provision exists for the eventual addition of a third. Each camera is on a separate hydraulically controlled support which can be rotated out of the way to allow access to the larger cameras at more remote positions on the north-south axis. The whole massive frame is supported at three points integral with the telescope piers and separated from the building structure, to avoid transmission of vibrations due to wind or to dome rotation.

The third floor includes a two-story chamber with an aluminizing tank capable of taking the 107-inch mirror, accessed from above and below through large hatches. A 20-ton crane is built into the dome roof structure, for handling this massive weight. It is also used to change the sections at the upper end of the tube which carry the various secondary mirrors.

Despite serious setbacks in the Westinghouse shops due in part to that company's commitments to nuclear reactor and submarine missile contracts, construction was accomplished without the enormous delays that had characterized the completion of many earlier large telescopes, including the 82-inch. The formal dedication ceremony of the 107-inch reflector took place on 26 November 1968, in the presence of a large and distinguished audience of university dignitaries, local residents, and astronomers from far and wide.

The transportation arrangements were quite different from those of 1939. By 1968, no one rode the railroads. On the day preceding the event, the VIP's were flown from Austin to Marfa in an Electra aircraft chartered from Braniff International Airways. The NSI's (Not So Important) came by chartered bus. The weather was beautiful and the VIP's were suitably "stayed with flagons" during the 400-mile trajectory.

Next morning was dismal. One of the present authors has a vivid recollection of debouching from the newly built Transient Quarters, a sort of motel for visiting astronomers, at six o'clock on the morning of the dedication. The brilliant clear sky had been replaced by torrential rain, which turned to heavy snow later in the day. Although the ceremonies in the dome, conducted by University of Texas President Norman Hackerman, passed off without incident, the weather made for a rather gloomy counterpoint to the stimulation of the event itself. One bright spot was provided during the proceedings when an old friend of Texas astronomy, engineer and industrialist Joe King, called for all those who had attended the 1939 dedication of the Otto Struve Memorial 82-inch reflector to stand. A handful of those present were able to share this distinction with him.

On their departure, the VIP's were loaded into two school buses for their trip to the Marfa Municipal Airport. As they embarked on the Electra, the weather looked increasingly threatening, with the rain now changing to snow even down at the airport. Once on board, the celebrants were served complimentary cocktails by attractive stewardesses wearing the rather provocative costumes then prescribed by Braniff, and the aircraft began rolling along the taxiway in the wake of a guiding vehicle. The passengers became aware of a bumpy motion, and then something flew up near the left wing. The pilot brought the aircraft to a stop with the remark "Ladies and gentlemen, we seem to have a problem." The main runway was sturdy enough to have been used as an emergency landing field for both commercial aircraft and World War II bombers, but the taxiway was not built to the same standards. The extremely heavy rain had undermined the pavement, and the four-engine Electra had broken through the surface with its port wheel assembly. A great arrow-shaped furrow some 25 yards long had been plowed in the taxiway, and the tip of the inboard port propeller was only about 4 inches above the ground ahead of it.

No serious damage was done to the aircraft, but it was clear that the VIP plane would go nowhere that day. Clutching their cocktails, the passengers transferred to two school buses, which nearly collided while driving off the airfield in the near zero visibility. The aircraft, in its Texas-colored livery of burnt orange, would sit there forlornly blanketed with snow for several days to come. Back in Marfa, the passengers were temporarily confined to their buses by a heavy hailstorm. Morale remained high, however, as J. J. (Jake) Pickle—a resourceful member of the U.S. Congress—collected enough ice in an inverted umbrella to freshen up the drinks.

Eventually the NSI bus appeared, having been hauled out of a soft patch on the road by a tractor. Amid taunts from the original passengers, a few VIP's were accorded space on the bus. It set off amid heavy snow on a dangerously icy road to Alpine, where a highway patrol officer warned against any attempt to take the northerly route via Fort Stockton. He told Anne Fleming, the Astronomy Department secretary, of the numerous trucks and cars ditched by the ice. Then he added, "That bunch of knuckle-heads who came for the dedication, their plane is grounded." She squelched him with the dignified reply: "We are the knuckle-heads." All hands safely reached Austin, with the snow area well past, in the small hours of the morning. Some of the VIP's took the Amtrak train from Marfa to San Antonio, beguiled en route by music provided by Representative Pickle, who was later pre-sented with a new harmonica in memory of the occasion.

The inauguration of the 107-inch telescope ushered in a new phase in the history of the McDonald Observatory. On the same mountain from which Struve, Greenstein, and the others had watched the V-2 rockets carry their spectrographs aloft in some of the first tentative steps toward space astronomy, there was now a major telescope designed as a ground-based support for NASA's ambitious program for exploring the solar system. The design had carefully insured, however, that it would be a general-purpose in-strument capable of profound studies in stellar and extragalac-tic astronomy as well, a valuable scientific resource for many decades into the future. Astronomical observations began in early March 1969, after Johnnie Floyd repaired the damage done by the incorrect installation of the drive gear by the Westing-house crew.

The new dome towers over Mr. McDonald's telescope a few hun-dred feet to the west, making the mountain just that much more visible from the Sierra Mulato south of the Río Grande, from the Glass Mountains to the southeast, and from the Sierra Viejo to the west. There are now two great white eggs, gleaming big and bright in the fierce West Texas sunlight. The banker's investment is ready to pay an even higher rate of dividend.

Epilogue

The entry into service of the 107-inch telescope completes the narrative of William Johnson McDonald's observatory. From this time forward, the telescope that he sponsored was no longer the primary instrument on Mount Locke, although the 82-inch reflector continues to give yeoman service and will do so well into the future.

A proper history of the McDonald Observatory in more recent times cannot yet be written, because we are too close to it to be objective. Thus, this last section is epilogue to the early history and prologue to the future. The 107-inch telescope has triggered an enormous increase in research activity at Mount Locke, and we do not wish that an account of the recent past negate our purpose. Consequently, we present only a few notes and updates from the past fifteen years, just to tidy things up and to show that McDonald's dream is not dissipating. We can assure, however, that both the scientific activities and the human foibles of the present and recent past will offer some future chronicler as much excitement and drama as we have found in the beginnings. We leave to him or her the multitudinous scientific discoveries, as well as the tales of the Swiss Army knife, the back-sliding fire truck, and many other charming vignettes.

Two tales must be told, however, one traumatic, the other quite warm. Firearms are very common in Texas. Astronomer Brian Warner had his tongue only slightly in cheek when he remarked that "Jeff Davis County is about the size of Israel—and slightly better armed." The prophecy inherent in the aphorism came to pass. The full precipitating causes may never be known, but one February night in 1970 a McDonald Observatory employee (not a Texan, but an Ohioan newly hired from another observatory!) suffered a breakdown and carried a pistol to the observing floor of the 107-inch telescope. He fired a shot at his supervisor, and then unloaded the rest of the clip into the primary mirror. Happily, fused silica is more resilient than ordinary glass, and the big mirror did not break. The craters have been bored out and painted black to reduce any light-scattering effect, and the end result is simply a slight reduction in the efficiency of the telescope. It is now the equivalent of a 106-inch telescope. The incident made the national television news, with Walter Cronkite describing it before a projection showing the wrong telescope upside down.

The retirement of Tommy and Eloisa Hartnett was in a totally different vein. Tommy reached the mandatory retirement age enforced by the University in 1977. His career had begun on the ranches of Jeff Davis County, proceeded to the construction site on Mount Locke, back to the ranches for a while, then to a permanent and valued position on the observatory staff. To the day of this writing, he bears the distinction, unshared with any other, of having served under every director, every resident astronomer, and every superintendent that the McDonald Observatory has ever known. His wife chose to retire with him, and the event was celebrated with the biggest barbecue in the history of the mountain. During the festivities, Harlan Smith presented the Hartnetts with a tooled leather photo album filled with pictures and letters from astronomers all over the world, in remembrance of their outstanding services to the observatory and to the astronomy that had been done there. It remains one of their most prized possessions. "Of course," Tommy says with a grin, "they made me cook the barbecue."

The quality of research at McDonald and at Austin has remained high, and the professional recognition has remounted in recent years, although not at the rate of the late 1940's; there is more competition now. R. Edward Nather and Brian Warner gained the Boyden Premium of the Franklin Institute of Philadelphia for work on the first optical pulsar and other high-speed variable stars. Gerard de Vaucouleurs' discovery of the local supercluster of galaxies, to which we belong, brought him the Herschel Medal of the British Royal Astronomical Society in 1980. Newcomer Donald E. Winget received the 1982 Trumpler Award of the Astronomical Society of the Pacific for the theoretical prediction and observational discovery of a new class of pulsating white dwarf stars.

In addition to these highly publicized prizes, there has been no lack of significant discoveries. Exhaustive studies of relative abundances of chemical isotopes in the atmospheres of stars, performed by David L. Lambert and his many associates, have provided critical tests of theories of stellar evolution. Thomas G. Barnes and David S. Evans invented a method of calculating star diameters from observations of their colors and apparent brightnesses, now called the "Barnes-Evans relation." Among the other discoveries made at McDonald in recent years: molecular oxygen in the atmosphere of Mars, methane ice on Pluto's surface, numerous molecular species in stellar atmospheres and the interstellar medium, many quasars nearly at the border of the observable Universe. The discoveries, however, are an exciting, but minor, part of the workaday world of the astronomer. As with other observatories, most of the work serves to fill in the gaps of existing knowledge. The McDonald telescopes have been used to gain insight into the development of supernovae, the formation and structure of galaxies, the phenomenon of starspots and star flares, the chemistry of comets, the dynamics of interstellar gas clouds, optical and physical properties of interstellar dust, the size and age of the Universe, and many other topics.

The McDonald Observatory has produced virtually all of the useful laser measures of distance to the lunar surface that exist today. The observatory was fortunate in securing the service of Eric C. Silverberg, long the only successful practitioner of lunar laser ranging. The original laser is now on exhibition at the Smithsonian Museum in Washington. As a byproduct of these studies, the axis intersection of the 107-inch telescope is probably the most accurately located point on the surface of the Earth; it currently serves as the fiducial point for a new high-technology geodetic coordinate system. While the original laser ranging system was not designed at McDonald, its location there is symptomatic of the fact that, since the arrival of Bob Tull and Ed Nather in the mid-1960's, the observatory has been in the forefront of the application of innovative optical and electronic technology to astronomical problems.

Observational expeditions to remote sites continue as the occasion demands. Two of them during the early 1970's, to an eclipse in Mauritania and to a very rare occultation of the bright star Beta Scorpii by Jupiter, were such massive undertakings that a departmental wag referred to the participants as "Harlan's Globetrotters."

This has been only a very incomplete listing of current activities. Perhaps the most remarkable thing about the McDonald Observatory today is the great diversity of work that is now *standard* in this place where Otto Struve defined such narrow limits in 1939. Future historians will have a harder time gathering all the threads together into a single strand.

The Observatory and the Department

The Department of Astronomy in the University of Texas at Austin materialized as a result of forces centered on the McDonald Observatory. Since the separation from Chicago, the department and the observatory have shared fully integrated offices. They have nonetheless remained administratively separate from one another, a separation necessitated both by university structure and by differences in funding. This fact was obscured from the casual bystander for a long while, since Harlan Smith held simultaneous appointments as observatory director and department chairman for so many years. Growth in both areas eventually made this an excessive burden, and in 1978 Paul A. Vanden Bout accepted the department chairmanship, to be succeeded by Frank N. Bash in 1982.

Prior to 1969, the distinction between department and observatory was not reflected very overtly in the activities of the astronomers. Excepting those who were getting the radio and millimeter wave programs underway, nearly all of the faculty astronomers observed at McDonald—and there were few astronomers on the McDonald staff who were not either university faculty or graduate students.

The 107-inch reflector, in addition to providing a powerful new observing tool, served as a harbinger of change—change in the types of astronomy to be done, change in the nature of the observatory, change in the relation with the department. The white monolith towering beside Mr. McDonald's mosque did not cause all of this, but signaled it.

One of the changes brought about by a plunge into NASA's space program was a need for more people, far more than the university could—or would—provide with lifetime positions on the faculty. Today, the McDonald Observatory staff numbers more nonfaculty professional astronomers than the department has faculty. Many of them teach an occasional course, and nearly all of them work with graduate students. Nonetheless, their existence underlines the fact that the observatory is a research institution with a separate life—and occasionally different goals—from the department.

On the other side, an increasing number of faculty astronomers no longer use the McDonald Observatory or related facilities. Part of this is due to the growing strength of theoretical astrophysics at Austin; as impressive and important as these studies are, they are essentially disconnected from the observatory itself and are thus beyond the scope of this history. Another factor is that aerospace technologies now permit exotic astronomy to be done from the upper atmosphere or from Earth orbit. More seriously, some find that McDonald now lacks a telescope adequate to maintain a frontline position in their fields of research; more about this later.

New Directions

In William J. McDonald's day, astronomy was restricted to the very narrow band of electromagnetic frequencies that we call visible light. Radiation from celestial objects is not so restricted, however. Radio signals from space were discovered in 1935. Since that time, astronomy has pushed into ever widening areas of observation. Even though optical astronomy is still the major single branch of the science, many astronomers associated with the McDonald Observatory do at least part of their research in areas unknown to the benefactor. Since these activities are in a sense extraneous to the subject of this history, brief summaries will serve for completeness.

Radio Astronomy

A few miles to the southeast of Mount Locke, beyond U.S. Highway 90, a flat patch of cattle range presents a curious sight. It looks as though someone is growing a bumper crop of helical shock absorbers. The 832 spirals, each about 10 feet long, are the receiving elements of the 2-mile interferometer of the University of Texas Radio Astronomy Observatory (UTRAO).

The sharpness of detail that any telescope can "see" depends on the ratio of its size to the wavelength of the radiation being collected. A wave of red light is nearly twice as long as a wave of blue light, so optical telescopes are less precise in red than in blue. The wavelength at which the 2-mile interferometer operates (near 1 meter) is some two million times longer than the middle of the window of visible light. To have the resolution capacity of even the smallest optical telescope on Mount Locke, the 30-inch reflector, a simple radio telescope at this wavelength would have to be nearly a thousand miles long.

Obviously, the telescope could not be that big, so it was necessary to make a non-simple radio telescope, using the principle of wave interference. If two receivers are separated by some distance, then the rotation of the Earth causes them to see the sky in slightly different ways. For radio signals from a given object, the phase of the wave detected with one antenna will shift, compared to what the other one detects. If the two signals are added together, there will be times when they are in phase and the signal is amplified. At other times, they will be out of phase, and the signals will cancel. The spacing and rapidity of these interference effects contain directional information.

The helices of the unique 2-mile interferometer are 832 complete radio antennae, arranged in a pattern roughly 2 miles square. The signals that they pick up from the sky are recorded on magnetic tapes. Untangling the astronomy from those signals uses a sizable fraction of the operational time of the giant central computer at the University of Texas at Austin.

In the early days of radio astronomy—and still, with some instruments—an observation could give the position of a radio source only very poorly. The uncertainties on a radio "position" might easily be as large as the entire field of an optical telescope, containing many visual objects. The primary work of UTRAO is to map the radio emission from the 80 percent of the sky that is observable from West Texas. The goal is to determine the positions of the 70,000 brightest radio sources with such high precision that they can be identified unambiguously with objects in the optical sky.

UTRAO is semi-independent. Its designer, builder, and director, James N. Douglas, reports directly to a university vice-president on some matters, but is responsible to the McDonald director on others. Frank Bash, who arrived shortly after Douglas in 1967, was also heavily involved in the design and construction of the telescope.

Millimeter Wave Astronomy

A quite different type of radio astronomy is now pursued at Mount Locke. Downslope to the southwest from the summit, below the access road that serves the two smaller optical telescopes, a rock bench jutting out of the mountainside supports yet another small dome. It and nearby sleeping quarters comprise the University of Texas Millimeter Wave Observatory (MWO).

Inside the dome is an equatorially mounted reflecting telescope with a 16-foot reflecting mirror providing photons to an f/0.5 prime focus detector, the very instrument whose dedication was witnessed by Harlan Smith on his interview trip to Austin in 1963. The mirror is not made of glass, but of metal. The reflective coating is not silver or aluminum, but gold. Instead of visible light, this telescope collects microwave radiation, whose wavelengths are about a thousand times longer than those of ordinary light. In the metric units commonly used throughout astronomy, the telescope collects wavelengths in the range 0.86 to 4.0 millimeters, hence the name of the facility. Millimeter radiation is on the high-energy, or short-wavelength, edge of the radio frequency spectrum.

Radiation with millimetric wavelengths characterizes relatively cool material, such as is found in the clouds of gas and dust that circulate in the inner regions of our galaxy. These molecular clouds are intimately involved in the formation of new stars. Millimeter wave studies of these clouds seek to map them, to study their potential for absorbing and polarizing light, and to detect the various species of molecule that they contain. The list of molecular species now known from this technique is increasing rapidly, now comprising nearly a hundred compounds. By a considerable extrapolation, the possibility is entertained by some astronomers that the simplest of the molecules essential for life may be detected in the future. So far, ethyl alcohol is among the more complex molecules detected; according to one's preference, that may or may not be included in the foregoing category.

Under the guidance of Paul Vanden Bout, now director of the National Radio Astronomy Observatory, the Millimeter Wave Observatory became a cooperative venture of the University of Texas at Austin's Departments of Astronomy and Electrical Engineering when the receiving dish was moved to Mount Locke from its earlier emplacement near Austin in 1967. In 1986 it is being moved to an even better site near the Arizona–New Mexico border, where its range can be extended to still shorter wavelengths.

Space Astronomy

Two of the fundamental reasons for doing astronomy in space reside in our own atmosphere. One is that, even under the best conditions, some fraction of the signals arriving at Earth is blocked out by the atmosphere. Those that do get through are buffeted around, and therefore distorted, by turbulence, temperature instabilities, and/or electromagnetic effects.

These facts explain why, historically, astronomers have chosen to work on isolated mountaintops. For most astronomy, higher is better, because it is closer to the top of the atmosphere. In many cases, though, even the highest mountains are too low. More detailed infrared work can be done from the upper atmosphere, because, for example, at 40,000 feet the blockage in that waveband is already down to only 0.1 percent of its sea level value. Even more sensitive infrared studies can be done from Earth satellites orbiting entirely outside the atmosphere. Observation of celestial X-ray and gamma ray sources can be done only from spacecraft or stratospheric balloons, because those wavelengths are almost entirely absorbed by the blanket of air.

Consequently, it was inevitable that McDonald staff and Austin faculty should not only do ground-based astronomy in support of space projects, but also become involved in observing from space and near-space. This has included the Mariner Mars orbiter, the Orbiting Astronomical Observatory, Uhuru, Copernicus, Einstein, the International Ultraviolet Explorer, Skylab, and NASA's Kuiper Airborne Observatory. Space Shuttle astronaut Karl Henize is an adjunct professor in the department. Four staff and three faculty were selected to Space Telescope science teams.

Otto Struve had been very quick to pounce upon promising new technologies, such as the Schmidt camera, Pyrex mirrors, and vacuum deposition of aluminum on mirror surfaces. He joined the space age at its beginning, with instruments on the V-2 rockets at White Sands. The recent space age has not caught Struve's successors resting on his laurels.

The Future

When it was built, the 82-inch Otto Struve reflector was the second largest telescope in the world. Thirty years later, the newer and bigger 107-inch reflector began its career as third largest. It now ranks about fifteenth. It won't get smaller, but it will continue to get "behinder." To remain among the world's great observatories, McDonald will have to think of—and get—a new instrument.

The skies of West Texas remain an undiminished asset. Despite occasional dust, passing storms, and the encroachment of some smog from Los Angeles, Mount Locke remains one of the prime astronomical sites in the United States. Cloudy weather rarely lasts long, and atmospheric scattering remains low. On the average, an astronomer has roughly a two to one chance of being able to use assigned telescope time for astronomical observation. Observing quality statistics are as notorious for overstatement as those encountered in the sale of used cars and horses, but sky darkness and transparency are as good at Mount Locke as anywhere in the continental United States. Satellite photographs of the United States at night show that the Davis Mountains area is one of the darkest in North America and almost unique in the combination of this feature with good weather, relative ease of accessibility, and existence of an observing complex.

McDonald Observatory made its reputation by its contributions to ground-based optical astronomy, using that term to denote all wavelengths accessible from the ground. Research from space and in the radio and millimeter wave bands does not make ground-based optical astronomy obsolete. Far from it. Space astronomy is enormously expensive and competition for access to instruments in space so keen that many worthwhile projects cannot be undertaken or are skimped for time. On the other hand, the infrared satellite IRAS, which surveyed almost the whole sky during its lifetime of only a few months, may have produced more new discoveries than any other instrument in the history of astronomy. Ground-based astronomers can take a longer-term approach to observation and can hope for extended runs at telescopes during which they may undertake some of the projects possible from the ground which have been squeezed off space instruments, or will seek optical counterparts of objects discovered in other wavebands from spacecraft.

Experience is dictating the need not for less ground-based astronomy, but for more, demanding larger apertures capable of detecting, especially with electronic aids, fainter and fainter objects, whether farther off at the limits of the Universe or intrinsically very faint such as planets circling nearby stars.

These are some of the arguments, proven well-founded, which stimulated the construction in the seventies and eighties of a series of new large telescopes in the 4-meter (150-inch) range in Arizona, Chile, Hawaii, Australia, the Canary Islands, and the Soviet Union's Caucasus Mountains.

In the eighties the push is for still larger instruments approaching the 10-meter (400-inch) range. Texas has not been backward in planning a telescope of similar size, of 300 inches diameter. It should be realized that the possible existence of even larger telescopes than this does not make one of this aperture instantly obsolete. The Universe offers an infinite variety of unsolved problems, and telescope productivity is just as much a function of the ingenuity of the observers, the sophistication of their detection equipment, and the quality of the observing site as it is of light grasp.

Such large telescopes cannot be simply scaled-up versions of previous instruments, and in particular their main mirrors (in several designs there are more than one) cannot be simply bigger, thicker, and heavier pieces of optical material. Entirely new engineering problems are encountered with increased size, and new methods are needed to solve them.

The Texas team has devoted several years of sophisticated design studies to the engineering problems of large telescope construction and has produced a design using a single large lightweight mirror, preferably of honeycomb design with a thin face supported on a complex system of ribs. The telescope tube will be arranged to turn about vertical and horizontal axes and will follow the slantwise motion of the stars by activators controlled by a small computer. The telescope house, including various laboratories and other rooms, will turn with the telescope.

All that is now needed is adequate funding, which is in the range of $50 million. This may seem a lot, but dollars are not what they were in William Johnson McDonald's day. His telescope kept McDonald in the forefront of research for half a century and is still an important instrument. The new "Eye of Texas" would do the same for a similar length of time.

Institutions such as the McDonald Observatory are rare and can only exist in favorable environments. One does not study glaciers in the Sahara or tropical flora in Alaska. Astronomy has found a highly favorable environment, both physical and social, in Texas. It has been a symbiotic relationship. Astronomy has a deep appeal to the sensibilities of the ordinary citizen, who feels the need to understand our place in this marvelous Universe. In addition, astronomy has a brilliant record both of advancement of pure science and long-term practical applications. The history of astronomy since William Johnson McDonald's time has shown that no one can predict how its development may change as the result of totally unexpected discoveries, but the astronomical heirs of William Johnson McDonald are deeply grateful for his legacy and hope to help it grow and flourish into the future.

APPENDIX A

The Last Will and Testament of William Johnson McDonald

State of Texas
County of Lamar

KNOW ALL MEN BY THESE PRESENTS:

I, W. J. McDonald of the above named State and County do hereby make, publish and declare this as my last will and Testament hereby revoking any other Will or Wills by me heretofore made.

* 1 *

I Give, devise and bequeath to Mrs. Florence McD. Rodgers, wife of A. N. Rodgers of Dallas, Texas, and to Mrs. Lillian McD. Brinton, wife of George D. Brinton of West Chester, Pennsylvania, my nieces, the sum of Fifteen Thousand ($15,000.00) Dollars each, and to William Stromeyer and his sister Irene Stromeyer, minors, children of the said Mrs. Lillian McD. Brinton by her former husband, the sum of Fifteen Thousand ($15,000.00) Dollars each.

I direct that my executors shall pay and satisfy the bequests contained in this item of my will free and discharged of any and all State or Federal inheritance tax, revenue taxes or taxes of any kind whatsoever.

* 2 *

I hereby appoint Mrs. Lillian McD. Brinton, trustee, without being required to give any bond, for the said William and Irene Stromeyer, to take possession of and handle, manage and control the bequests herein given them until they shall respectively arrive at the age of twenty-one years, or Irene shall marry, upon the happening of which event said trustee shall pay over to them their respective legacies, that is to say, such payment shall be made to the said William Stromeyer when he arrives at the age of twenty-one years and to the said Irene when she shall either arrive at the age of twenty-one years, or upon her marrying before attaining the age of twenty-one years. In case of death of either William or Irene during minority the bequest herein to the one dying shall go to the survivor, *provided* that in the case of William if he should so die being married at that time, then his bequest shall descend and vest in his heirs in accordance with the laws of the State of Texas; and in the case of Irene if she should so die being married at the time, then her bequest being disposable by her and her own will shall be subject to such disposition or in case of intestacy to the laws of descent of the place governing it.

* 3 *

I give, devise and bequeath to my nephew, Charles McDonald for and during his life only, Fifteen Thousand ($15,000.00) Dollars in United States Treasury Four Per Cent bonds, subject to the trust herein contained.

I direct that my executors shall pay and satisfy the bequest contained in this item of my will free and discharged of any and all State or Federal inheritance tax, revenue taxes or taxes of any kind whatsoever.

I hereby appoint the First National Bank of Clarksville, Clarksville, Texas, trustee for the said Charles McDonald, without being required to give any bond, to receive the bequest herein given to him, and said trustee shall register said bonds in its name as such trustee, and I hereby direct that the said trustee pay to the said Charles McDonald the net revenue arising from the interest on said bonds, as and when received, during the life of the said Charles McDonald, and at his death this bequest shall go, share and share alike to his children, James, William and Rosemary McDonald, provided that should the said Charles McDonald die during the minority of said children or anyone of them, this trust shall continue in force as to said bonds for the use and benefit of such minor children or child during their or its minority, and upon the arrival at the age of twenty-one years, respectively, each shall receive his or her share, or should the said Rosemary marry before arriving at the age of twenty-one years, then upon the happening of that event she shall receive her share.

During the minority of said children or anyone of them, after the death of the said Charles McDonald, should he die during such minority, the trustee shall use the net interest from the share of such minor children or child as the case may be, for the same purpose and in the same manner as is hereinbefore [*sic*] provided for in Item Four of this Will for the use of the interest on the bonds there devised to such children.

* 4 *

I give, devise and bequeath to the said James, William and Rosemary McDonald, children of Charles McDonald, each Fifteen Thousand ($15,000.00) Dollars, in United States Treasury Four Per Cent Bonds, subject to the trust herein contained.

I direct that my executor shall pay and satisfy the bequests contained in this item of my will free and discharged of any and all State or Federal inheritance tax, revenue taxes or taxes of any kind whatsoever.

I hereby appoint the First National Bank of Clarksville, Clarksville, Texas, trustee for the said James, William and Rosemary McDonald, without being required to give any bond to receive the bequests herein given to them, and said trustee shall register said bonds in its own name as such trustee, and I hereby direct that the said trustee shall use the net interest from said bonds, or so much thereof as may be necessary, in the support, education and maintenance of said children, and upon the arrival of each of them, respectively at the age of twenty-one years, each shall receive the amount coming to him or her under this bequest, or should the said Rosemary marry during her minority, then upon the happening of that even she shall receive the amount coming to her under this bequest.

If any of said children shall die without issue, and before arriving at the age of twenty-one years, the legacy of this one so dying, both that devised in this item of my will and that which may have accrued upon the death of the said Charles McDonald, shall go to the survivor or survivors of them, and if all shall die before arriving at the age of twenty-one years and without issue, then all the interest or share in my estate by this will devised to them shall go to my residuary legatees.

* 5 *

All the rest, residue and remainder of my estate, I give, devise and bequeath to the Regents of the University of Texas, in trust, to be used and devoted by said Regents for the purpose of aiding in erecting and equipping an Astronomical Observatory to be kept and used in connection with and as a part of the University for the study and promotion of the study of Astronomical Science. This bequest is to be known as the W. J. McDonald Observatory Fund. As soon as my Executors shall have paid off all charges against my estate, settled with all legatees and have the estate in shape to turn over the residue to the Regents, they shall do so using only such time to accomplish this as in their judgment is necessary, reasonable and proper. Upon receipt of such residue the Regents shall proceed at such time and manner as in their judgment may seem best to execute the trust and they shall have full power and authority to administer and handle the same for the purpose of carrying out its purpose and object, and in so doing they may apply all or any part of the income from the bequest and all or any part of the corpus of the bequest as they may deem proper to use towards such erecting and equipping the Observatory, it being my intention that the Regents shall have full power and authority to handle, use and appropriate this bequest, corpus and all accrued income in such manner as to them may seem best in order to carry out the object of the bequest, the only limitation on their authority and power being that the bequest is intended solely for the use and benefit of an Astronomical Observatory in one or all the ways hereinbefore mentioned. The Regents shall have power to sell any real estate at any time.

All investments are to be made in such bonds and securities as are prescribed by law for the investment of the State Common School Fund.

In handling the notes due my estate the Regents are requested to use the utmost liberality and leniency in the matter of renewals and extensions consistent with safety.

* 6 *

My executors are directed to expend the sum of Fifteen Hundred ($1500.00) Dollars to build in the Cemetery lot of J. T. McDonald in Evergreen Cemetery, a family monument to be as near as may be in general style and outline like my mother's monument in the Old Grave Yard at Paris, Texas.

* 7 *

I hereby appoint Morris Fleming of Paris, Texas, and the First National Bank of Clarksville, Clarksville, Texas, Executors of this Will and specially direct that no bond be required of them as such, and I further direct that no action shall be had in the Probate Courts in relation to my estate than the probating of this Will and the return of an Inventory and list of claims of my estate.

In case of the death, refusal or inability to act of either of said named Executors, the survivor shall have and exercise all authority hereunder as sole executor.

* 8 *

This Will is executed by me in duplicate originals, each to be of equal validity, but made in duplicate as a precaution against one becoming lost, and upon the filing and probating of either one, the other is to have no further force and effect than corroborative, and the one so filed and probated is to be the Will.

* 9 *

For purposes of identification I have signed my name on the left margin of each sheet of this Will.

* 10 *

In testimony whereof, I, the said W. J. McDonald have this 8th day of May A.D., 1925, signed my name hereto and on the said margins, and we, whose names appear hereto as subscribing witnesses have on the same day, and at the request and in the presence of the testator and in the presence of each other signed our names hereto as such subscribing witnesses.

W. J. McDONALD

Subscribing Witnesses:
  H. L. Baker
  J. M. Caviness

APPENDIX B

The Directors of the McDonald Observatory

Otto Struve	Nov 1932–Aug 1947
Gerard P. Kuiper	Sep 1947–Dec 1949
Otto Struve	Jan 1950–Jun 1950 [1]
Subrahmanyan Chandrasekhar	Jul 1950–Dec 1950 [2]
Bengt Strömgren	Jan 1951–Aug 1957
Gerard P. Kuiper	Sep 1957–Mar 1959 [3]
William W. Morgan	Apr 1959–Aug 1963
Harlan J. Smith	Sep 1963–

Notes

1. Struve was not appointed formally as director during this period. His official title was Honorary Director and Chairman of the Observatory Council, but he exercised the authority in the absence of a formal director, following Kuiper's resignation.

2. Chandrasekhar was never appointed formally as director. His official title was Acting Chairman of the Astronomy Department, but he exercised directorial authority after the departure of Struve and until Strömgren could take up the post.

3. Most sources place this transfer of power in 1960, but the Yerkes archives make it clear that the above-cited dates are correct. The confusion arises from the fact that Kuiper retained the chairmanship of the Chicago department until early 1960, so that he could complete the merger of the Chicago and Austin departments.

The Telescopes of the University of Texas Observatories

The McDonald Observatory at Mount Locke

107-inch (2.7-meter) reflector, Westinghouse, 1969
82-inch (2.1-meter) Otto Struve Memorial reflector, Warner and Swasey, 1939
36-inch (0.91-meter) reflector, Boller and Chivens, 1956
30-inch (0.76-meter) reflector, Boller and Chivens, 1970
30-inch (0.76-meter) McDonald Laser Ranging Station, 1981
12-inch (0.30-meter) Transportable Laser Ranging Station, 1979
    This instrument is vehicle-mounted and operates in the field.
3-inch (0.08-meter) fisheye patrol camera, 1964

The Millimeter Wave Observatory (Mount Locke)

16-foot (4.9-meter) dish antenna, 1963
    This instrument was located near Austin until 1967 and may be transferred to Mount Graham, New Mexico, in about 1986 with improved capability in very much shorter wavelengths.

The Radio Astronomy Observatory (Marfa)

2-mile linear interferometer, 1975

References

The authors have provided an extensive bibliography of the many hundreds of sources that went into this history to the McDonald Observatory archives. Only those items required to support direct quotations and other important points are presented here. Sources for archival documents are indicated according to the following codes:

BTHC   Barker Texas History Center, University of Texas at Austin
CWR   Freiburger Library, Case Western Reserve University
McD   McDonald Observatory archives
YO   Yerkes Observatory archives

1. G. Grubb, McDonald Observatory employee, interview by JDM, 8 November 1980.

2. F. Kahl, Manager, Marfa (Tex.) Municipal Airport, interview by JDM, 22 July 1983.

3. This section is a synthesis of many sources, most of them articles from various Texas newspapers during the period just after McDonald's death, during the legal proceedings, and in early May 1939. Detailed referencing is impractical, except for the few places where it seems important, such as verbatim citations. (BTHC, McD)

4. S. Acheson, "Donor Rests in Obscure Grave," *Dallas News,* 30 April 1939.

5. Books from W. J. McDonald's personal library are on display in the entrance lobby to the dome of the 82-inch telescope at Mt. Locke.

6. W. J. McDonald II, interview by DSE, 1980.

7. C. Pollard et al., Brief and Argument for Appellees in the Case of Mrs. Florence Rodgers et al. vs. Morris Fleming et al., Court of Civil Appeals for the Sixth Supreme Judicial District of Texas, Texarkana, 1926. (BTHC)

8. H. Y. Benedict, letter to T. J. J. See, 1 May 1926. (BTHC)

9. H. Y. Benedict, letter to E. B. Frost, 27 March 1926. (YO)

10. E. B. Frost, letter to H. Y. Benedict, 1 April 1926. (YO)

11. The accounts of the trial given in this section are synthesized largely from reference 7 above and from day-by-day reports published in

the *Paris* (Tex.) *Morning News* in August–September 1926. All direct quotations are from reference 7 unless otherwise noted.

12. H. Y. Benedict, letter to U. Texas lawyer T. L. Beauchamp, 9 August 1928. (BTHC)

13. *Paris* (Tex.) *Morning News* (September 1926).

14. R. B. Levy, Judgment in Case No. 3353, Mrs. Florence Rodgers et al. vs. Morris Fleming et al., Court of Civil Appeals for the Sixth Supreme Judicial District of Texas, Texarkana, 21 April 1927. (BTHC)

15. O. Speer, Judgment in Case No. 846-4923, Commission of Appeals, Section B, Supreme Court of the State of Texas, 29 February 1928. (BTHC)

16. As this was not a trial, but a mistrial, there is no official court record. The major source for this section, including all quotations, is the *Paris* (Tex.) *Morning News*, in its numbers of 30 October–23 November 1928.

17. E. B. Frost, letter to U. Chicago president Max Mason, 2 February 1928. (YO)

18. E. B. Frost, letter to M. Mason, 5 April 1928. (YO)

19. John T. Moulds, letter to E. B. Frost, 16 April 1928. (YO)

20. E. B. Frost, letter to H. Y. Benedict, 25 February 1926.(YO)

21. The earliest official reference that we have found to a possible Texas-Chicago collaboration is a letter from "Frances M. Little, secretary to the President of the University of Texas, dated March 22," cited by U. Chicago president R. M. Hutchins, memorandum to G. O. Fairweather, 30 May 1932. (YO)

22. P. Vandervoort, Yerkes Observatory astronomer, interviewed by JDM, 13 January 1981.

23. W. W. Morgan, Yerkes Observatory astronomer, interviewed by JDM, 15 March 1982. Morgan's version of the Kuehne visit is supported implicitly by reference 32 below.

24. Warner Seely, internal Warner and Swasey Company memorandum to C. J. Stilwell, 22 October 1932. (CWR)

25. O. Struve, letter to Dean H. G. Gale, 30 April 1932. (YO)

26. Note in Otto Struve's hand, headed "telephoned to Chicago," 11 May 1932. (YO)

27. O. Struve, letter to R. M. Hutchins, 21 May 1932. (YO)

28. "The McDonald Observatory Agreement between the University of Texas and the University of Chicago," 23 November 1932. (YO, BTHC)

29. Many identical letters, e.g., O. Struve to H. N. Russell, 10 May 1932. (YO)

30. O. Struve, letter to H. G. Gale, 5 May 1932. (YO)

31. H. G. Gale, letter to O. Struve, 7 May 1932. (YO)

32. F. Schlesinger, letter to O. Struve, 22 July 1932. (YO)

33. H. Shapley, letter to O. Struve, 14 May 1932. (YO)

34. T. J. J. See, letter to H. Y. Benedict, 17 September 1932. (BTHC)

35. O. Struve, "The Birth of the McDonald Observatory," *Sky & Telescope Magazine* 24, no. 6 (December 1962): 318.

36. *Ciel et Terre Magazine,* April 1934.

37. The earliest record of the Warner and Swasey Company's involvement in the McDonald bequest predates that of Yerkes. W&S Texas field representative L. M. Cole telephoned the company on 2 February 1932 concerning the university's plans for a telescope, according to a memorandum from company Secretary W. Seely to then Vice-President C. J. Stilwell, 22 October 1932. (CWR)

38. O. Struve, letters to H. G. Gale, 18 March and 30 April 1932. (YO)

39. O. Struve, letter to Robt. L. Holliday, undated but among papers from late 1932 (YO). This letter is important, because there is much controversy among current McDonald and Fort Davis people concerning the naming of Mount Locke. We cite from this letter: ". . . Flattop Mountain (which some insist on calling Mount Locke, and others prefer to call U Up and Down Mountain) . . ." All three names already existed, and the minor peak to the east (now called Flattop) was then called variously Little Flattop or Mount Fowlkes.

40. Keesey Miller, Fort Davis resident, interview by JDM, 27 December 1980.

41. W. W. Negley, letter to H. Y. Benedict, (no day) January 1933. (YO)

42. W. S. Miller, Fort Davis resident, letter to O. Struve, 29 May 1933 (year mistyped as 1931). (YO)

43. Contract between the Regents of the University of Texas and the Warner and Swasey Company (clause IVe), October 1933. (BTHC)

44. O. Struve, "Memorandum Concerning Mr. Van Biesbroeck's trip to the Davis Mountains, April, 1933," Yerkes Observatory internal document, undated. (YO)

45. G. van Biesbroeck, letter to O. Struve, 7 April 1933. (YO)

46. D. B. Dumas, H. J. Dorman, and G. V. Latham, *Bulletin of the Seismological Society of America* 70 (1980): 1171.

47. G. van Biesbroeck, letter to O. Struve, 28 November 1933. (YO)

48. B. Scobee, "Building of McDonald Observatory, II," *West Texas Historical and Scientific Society, Publ. No.* 7 (Alpine, Tex.: Sul Ross State Teachers College, 1937), p. 45.

49. *Christian Science Monitor,* 8 March 1935.

50. G. van Biesbroeck, letter to O. Struve, 17 April 1933. (YO)

51. G. van Biesbroeck, letter to O. Struve, 13 November 1933. (YO)

52. O. Struve, "Research Program of the Yerkes and the McDonald Observatories, *Popular Astronomy* 41 (1933): 543.

53. In fact, Struve's penchant for amalgamation of the two observato-

ries is today a serious hindrance to unraveling what happened at one place or the other. The annual reports published in the *Astronomical Journal* make no distinction and frequently do not even mention the instruments used for specific work. Often, separation is possible only by concordance of the annual reports with the archival correspondence files.

54. O. Struve, letter to H. Y. Benedict, 13 November 1935. (YO)

55. F. E. Roach, "The Early Days at the McDonald Observatory," reminiscences of 1934–1936, 5 May 1980.

56. O. Struve, letter to C. T. Elvey, 7 April 1936. (McD)

57. O. Struve, letter to C. T. Elvey, 25 June 1936. (McD)

58. O. Struve, letter to C. T. Elvey, 16 February 1937. (McD)

59. C. T. Elvey, letter to O. Struve, 1 March 1937. (McD)

60. O. Struve, letter to C. T. Elvey, 11 March 1937. (McD)

61. O. Struve, letter to C. T. Elvey, 6 November 1937. (McD)

62. B. Jester, letter to R. M. Hutchins, 12 February 1934. (YO)

63. W. Seely, letter to O. Struve, 30 July 1932. (YO) This letter, as well as the Annex to reference 28, negate the local folklore (L. Hobbs, private communication) among current Yerkes astronomers that the coudé configuration was a last-minute afterthought.

64. Nearly all of our information on the fabrication of the 82-inch mirror has been obtained from the archives of the McDonald and Yerkes Observatories or from private sources. The relevant material has been expunged from the Warner and Swasey archives, now residing in the Freiberger Library of the Case Western Reserve University. In particular, those archives lack the voluminous correspondence about the 82-inch mirror, and the personnel file marked C. A. R. Lundin is empty, save the initial announcement of his affiliation with the company. Details of working the mirror are taken directly from Lundin's handwritten workbooks, made available to us by his daughter, Ruth Douglas.

65. O. Struve, "The Story of an Observatory," *Popular Astronomy* 55 (1947): 283.

66. These figures are from Lundin's workbooks. Our own calculations differ slightly from these values.

67. Despite a formal inquiry to the Warner and Swasey Company, we have been unable to ascertain the eventual fate of the 60-inch flat mirror used in testing the 82-inch primary. It is surprising that such a large piece of glass, a valuable engineering resource, could disappear with no trace. It may have gone to Córdoba, Argentina.

68. O. Struve, letter to C. T. Elvey, 8 September 1936. (McD)

69. J. S. Plaskett, letter to Warner and Swasey President P. E. Bliss, 6 May 1933. (CWR)

70. O. Struve, letter to C. T. Elvey, 10 May 1937. (McD)

71. O. Struve, letter to C. T. Elvey, 19 October 1937. (McD)

72. O. Struve, letter to C. T. Elvey, 3 December 1937. (McD)

73. O. Struve, letter to C. T. Elvey, 31 March 1938 (McD). This letter, elegantly written in Struve's hand, includes explanatory diagrams.

74. O. Struve, letter to C. T. Elvey, 5 May 1938; letter to U. Texas President J. W. Calhoun, 10 May 1938; letter to C. T. Elvey, 19 May 1938. (McD)

75. L. J. Peter and R. Hull, *The Peter Principle* (New York: Morrow, 1969). For those not familiar with this seriocomic analysis of corporate management, the reference is to the common practice of promoting personnel from levels where they are very competent to higher ones where they are not.

76. J. Texereau, French astronomer, letter to DSE, 28 April 1980; interview (in French) by JDM, 24 September 1980.

77. J. S. Plaskett, "Report on Installation and Testing of Optical Parts of McDonald Telescope," document undated, but annotated as received 20 March 1939, apparently by C. J. Stilwell. (CWR)

78. C. J. Stilwell, letter to O. Struve, 2 June 1939. (YO)

79. C. J. Stilwell, letter to J. S. Plaskett, 7 October 1940. (CWR)

80. O. Struve, letter to C. T. Elvey, 11 March 1939. (McD)

81. C. Richmond, letter to C. T. Elvey, undated but obviously April 1939, based on content and its location in the archives. (McD)

82. J. C. Kline, letter to W. Seely, undated but obviously early 1939. (McD)

83. Bart Bok, Harvard astronomer, private conversation with JDM, 9 June 1982.

84. "Addresses Made at the Dedication Exercises of the W. J. McDonald Observatory," Warner and Swasey Company, 1939. (McD)

85. J. L. Greenstein, CalTech astronomer, telephone interview by JDM, 9 August 1983.

86. S. Chandrasekhar, telephone conversation with JDM, 31 January 1986.

87. M. J. Schwarzschild, Princeton astronomer, private conversation with JDM, 9 June 1982.

88. These telegrams and typescripts are in the McDonald archives.

89. C. T. Elvey, letter to J. W. Calhoun, 26 August 1939. (McD)

90. U. Chicago accident report, 23 July 1939. (McD)

91. T. L. Page, NASA astronomer, interview by JDM, 13 January 1981.

92. C. T. Elvey, letter to C. K. Seyfert, 28 July 1939. (McD)

93. Pol Swings' given name has been spelled many ways, and even Swings himself did not know what was correct. He explained to the University of Chicago Personnel Office that his parents and the official documents relating to his birth alternate between Polidor and Polydor, but that he had abandoned the problem when he was yet young, adopting the uncontested abbreviation Pol.

94. T. Immega, letter to F. E. Roach, 29 August 1950 (copy kindly provided to us by Dr. Roach).

95. R. M. Hutchins, letter to O. Struve, 11 November 1939; O. Struve, letter to R. M. Hutchins, 13 November 1939 (YO). Our attention was brought to this curious exchange by David DeVorkin, National Air and Space Museum.

96. R. M. Hutchins, letter to H. P. Rainey, 9 April 1940; O. Struve, "Plan for Astronomical Collaboration in Connection with the McDonald Observatory," unpublished, 16 March 1940. (YO)

97. O. Struve, "Annual Report of the Yerkes and McDonald Observatories for 1941–1942," *Transactions of the American Astronomical Society* 10 (1942): 289.

98. In fact, in later years Kuiper told his Fort Davis friends, the Millers, that he had worked with the Office of Strategic Services (predecessor to the Central Intelligence Agency) part of the time. Struve gives circumstantial support to this idea by commenting that Kuiper's wartime activities sometimes involved "personal danger to himself," and Mrs. Kuiper's letters to the McDonald staff during the war attested that he was often away on secret business. Whatever the truth here, Kuiper definitely was a member of the ALSOS group, which went to Germany immediately after the surrender to investigate the Nazi nuclear weapons program; Aden Meinel participated in this activity also. A minor side result of this was a confidential memorandum that Kuiper circulated to American observatories in 1946, detailing the degree of complicity by German astronomers with the Hitlerian regime.

99. D. H. DeVorkin, "The Maintenance of a Scientific Institution: Otto Struve, the Yerkes Observatory and Its Optical Bureau during the Second World War," *Minerva* 18, no. 4 (1980): 595.

100. T. Harnett, interview by JDM, 23 December 1980.

101. S. Chandrasekhar, interview by JDM, 14 March 1984.

102. Our only source for this is a newspaper clipping, with neither the date nor the paper identified, in the Yerkes archives.

103. G. P. Kuiper, "Annual Report of the Yerkes Observatory and the McDonald Observatory for 1948–49," *Astronomical Journal* 54 (1949): 223.

104. M. Krebs, letter to DSE, 5 November 1981.

105. The Yerkes archives of the period 1 March 1959–1 March 1960 present a confused picture, but it is clear that Morgan was completely in charge at least from late 1959 onward. For example, in February 1960, Morgan told the Austin administration that Kuiper had not even been consulted on the 1960 budget.

106. H. J. Smith, verbal reminiscences, 1984.

107. J. Texereau, "Refiguring the 82-inch McDonald Reflector's Optics," *Sky and Telescope Magazine* 28 (December 1964): 345.

Index